Underwater Vehicle Control and Communication Systems Based on Machine Learning Techniques

The development of intelligent transportation systems, especially autonomous underwater vehicles, has become significant in marine engineering, with an aim to enhance energy efficiency management and communication systems. This book covers different aspects of optimization of autonomous underwater vehicles and their propulsion systems via machine learning techniques. It further analyses hydrodynamic characteristics including the study of experimental investigation combined with hydrodynamic characteristics backed by MATLAB® codes and simulation study results.

Features:

- Covers utilization of machine learning techniques with a focus on marine science and ocean engineering.
- Details effect of the intelligent transportation system (ITS) into the sustainable environment and ecology system.
- Evaluates performance of particle swarm intelligence-based optimization techniques.
- Reviews propulsion performance of the remote-controlled vehicles based on machine learning techniques.
- Includes MATLAB® examples and simulation study results.

This book is aimed at graduate students and researchers in marine engineering and technology, computer science, and control system engineering.

Underwater Vehicle Control and Communication Systems Based on Machine Learning Techniques

Edited by

Tien Anh Tran, Kolla Bhanu Prakash,
Subrata Chowdhury and Ivan CK Tam

CRC Press
Taylor & Francis Group
Boca Raton London New York

CRC Press is an imprint of the
Taylor & Francis Group, an **informa** business

Designed cover image: © Tien Anh Tran, Kolla Bhanu Prakash, Subrata Chowdhury and Ivan CK Tam

First edition published 2024
by CRC Press
2385 NW Executive Center Drive, Suite 320, Boca Raton FL 33431

and by CRC Press
4 Park Square, Milton Park, Abingdon, Oxon, OX14 4RN

CRC Press is an imprint of Taylor & Francis Group, LLC

ISBN: 9781032335346 (hbk)
ISBN: 9781032335353 (pbk)
ISBN: 9781003320074 (ebk)

DOI: 10.1201/9781003320074

Typeset in Times
by Newgen Publishing UK

Contents

About the Editors

Tien Anh Tran (IEEE Member, IMarEST Member) is a research fellow at Seoul National University, Seoul city, South Korea.

He is an assistant professor (lecturer) at the Department of Marine Engineering, Vietnam Maritime University, Haiphong City, Vietnam. Additionally, he is also an honorary professor at the School of Computing Science and Engineering, Galgotias University, India. He received an BEng and MSc at Vietnam Maritime University, Vietnam in 2011 and 2014, respectively. He got a PhD. Degree at Wuhan University of Technology, Wuhan, China in 2018. He is an editor/guest editor for the reputation journals indexed in SCI/SCIE. In 2015, he had been awarded the Chinese Government Scholarship (CSC) for the full funding of the Doctor of Philosophy (PhD) program in China. In 2019, he was awarded the NEPTUNE prize for the outstanding researchers by Vietnam Maritime University. In 2022, he was one of five outstanding scientists in Vietnam to be nominated for Ta Quang Buu prize by the National Foundation for Science & Technology Development (NAFOSTED). Additionally, he was selected and awarded the full scholarship for the Postdoctoral Fellowship Program of National Research Foundation (NRF) for Foreign Researchers by the Government of South Korea.

Kolla Bhanu Prakash is working as Professor and Research Group Head for A.I & Data Science Research group at KLEF. He has published 100 research papers in international, national journals and conferences. He has published Scopus and SCI publications of 82 with H-Index of 17 and the total citations are 940. He is a reviewer for several international journals and has served as a reviewer, keynote speaker and TPC member for several international conferences. His research interests include deep learning, data science and quantum computing. He has received a Best Researcher Award, has authored 12 books with international reputed publishers and also published 9 patents and 1 copyright. He is guiding 4 PhD scholars, and 2 scholars got awarded under his guidance.

Subrata Chowdhury is completing an MTech at the Sreenivasa Institute of Technology and Management Studies, Chittoor Andhra Pradesh, India. He has edited 5 books in association with the CRC press and others. His research areas include data mining, big data, machine learning, quantum computing, fuzzy logic, AI, edge computing, Swarm intelligence, and healthcare. He has received awards and nominations from the different national and international science societies. He had published more than 50 papers in international and reputed journals. He has been a member of the IET and a member of the IEEE. He is also a reviewer for the journals. He has been invited as a keynote speaker for many workshops, conferences and seminars.

Ivan Tam is an associate professor at Newcastle University in the UK, with a wealth of experience in multi-disciplinary research and a strong track record of leading innovative projects. His research interests are in the clean fuel combustion process, exhaust emission control, energy management and renewable energy technology. He has successfully applied cryogenic technology and organic Rankine cycles in improvement of performance in marine engines. He is currently developing a novel wet scrubber technology in the mitigation of gas emissions including carbon dioxide with his research team. He has a good connection with engineering companies through the network of learned societies and consultancy work with local industries in Singapore. He is a Fellow member, Assessor of Registration, the Honorary Advisor and Immediate Past Chairman of the Joint Branch (IMarEST & RINA). He received the President's Commendation Award from IMarEST in 2016. He is the Director of Degree Programmes in Marine Technology (Marine Engineering, Naval Architecture & Offshore Engineering) at the university. He is currently holding an appointment as the Director of Innovation & Engagement. A key aim of this new office is to increase engagement with alumni and partnership with industry.

Preface

The transportation systems play an important role in the economic development of nations through carried cargo between countries. The development of intelligent transportation systems (ITS) becomes more and more significant in shipping transportation with the aim of decreasing maritime accidents and risk management along with ship energy efficiency management and communication systems. The contribution of research plays a vital role in R&D of devices and autonomous underwater vehicles (AUV) and communication systems. The utilization of information technology (IT) in intelligent transportation systems (ITS) is ongoing. Especially, the development of Industry 4.0 is a big milestone in science and technology. Machine learning techniques are efficient tools to predict and evaluate working performance and reliability of uncertain motion and nonlinear dynamic systems in ocean engineering. Therefore, research into machine learning techniques in these specific fields is of high importance currently.

The development of science and technology always brings more and more benefits for all majors, in particular ocean engineering. The prediction and development of intelligent models are becoming popular for utilizing achievement from the development of science and technology. Underwater vehicles and communication systems are uncertain objects that always are influenced by many external factors from the environment. Therefore, there are many aspects approaching this field through different methods. Moreover, the underwater vehicles are tracking variably, including the experimental investigation and the state-of-the-art methodologies to improve controlling quality of these devices. Additionally, there are a series of related fields in ocean engineering, including safety science, environmental science, risk management, modern control theory, modelling and simulation, and optimization engineering. This topical collection book has been proposed to solve related fields by applying machine learning techniques in ocean engineering and its applications. The autonomous underwater vehicles and communication systems normally operate in the field of ocean engineering and operate at deep ocean levels to help warn of dangers in hazardous environments. Moreover, autonomous underwater vehicles related to optimization of their propulsion systems during preliminary design along with analyzing hydrodynamic characteristics.

The study of experimental investigation combined with hydrodynamic characteristics will be conducted and presented in this book through some different research results. The studies of unmanned vehicles are meeting a lot of difficulties including both remote control engineering and hydrodynamic characteristics of autonomous underwater vehicles (AUV). Theoretical and experimental investigation will be employed to research autonomous underwater vehicles through their different behaviors. The development of machine learning techniques is becoming more common in the field of marine engineering. Therefore, the utilization of machine learning techniques would be conducted in this book with a focus on marine science and ocean engineering. The R&D will be expressed in this book through fundamental state of the art results in the amalgamation of computer science and ocean engineering.

The accepted chapters have been selected and evaluated carefully through the blind-peer review processing by the experts in ocean engineering, computer science, and control system engineering. All selected chapters are the high-quality research results from the contributors who come from countries in the world such as Peru, India, China, South Korea, Singapore, Thailand, Vietnam and so on. These contributions have made the interesting book "*Underwater Vehicle Control and Communication Systems Based on Machine Learning Techniques*". Lastly, we highly appreciate the great contributions from authors/editors and editorial team of CRC Press Taylor & Francis to work hard for publishing this book.

Editors
Tien Anh Tran, PhD.
Kolla Bhanu Prakash, PhD.
Subrata Chowdhury, PhD.
Ivan CK Tam, PhD.

Contributors

B. Akoramurthy
Reva University, Bangalore, India

A. Bharathi Anbarasan
Dr. N. G. P Arts and Science College, Coimbatore, India

G. Arunkumar
Department of CSE, Madanapalle Institute of Technology & Science, Madanapalle, India

Mingzhi Chen
Shanghai Maritime University, Shanghai, China

William Dextre-Martinez
Universidad Nacional Santiago Antúnez de Mayolo, Huaraz, Peru

J. Dhilipan
Department of Computer Application, SRMIST, Tamapuram, Chennai, India

K. Dhivya
Department of CSE, Pondicherry University, Pondicherry, India

Manoj Kumar Gupta
School of CSE, Shri Mata Vaishno Devi University, Katra, India

Ashish Kumar
Department of CSE, ITS Engineering College, Greater Noida, India

Brahmadesam Viswanathan Krishna
Department of CSE, Rajalakshmi Engineering College, Chennai, India

Hongfei Li
Shanghai Maritime University, Shanghai, China

B. Akora Murthy
Sri Venkateswara College of Engineering and Technology, India

S. Prabavathy
Arulmigu Subramania Swamy Arts and Science College, India

Edwin Ramirez-Asis
Universidad Señor de Sipán, Chiclayo, Peru

K.S. Riya
Department of CSE, Veltech Multitech, Dr Rangarajan Dr Sakunthala Engineering College, Chennai, India

D. Saravanan
Department of CSE, IFET College of Engineering, Villupuram, India

G. Vennira Selvi
Sri Venkateswara College of Engineering and Technology, Chennai, India

Ashish Seth
School of Global Convergence Studies, INHA University, South Korea

Kshatrapal Singh
School of CSE, Krishna Engineering College, Ghaziabad, India

Haoxuan Song
Shanghai Maritime University, Shanghai, China

Suman Kumar Swarnkar
Department of Computer Science and Engineering, Shri Shankaracharya Institute of Professional Management and Technology, Raipur, Chhattisgarh, India

M. Vedhapriya
Department of Computer Science, SRMIST, Tamapuram, Chennai, India

Tien Anh Tran
Department of Marine Engineering, Vietnam Maritime University Haiphong, Vietnam

1 A Survey on Enhancement and Restoration of Underwater Image

Challenges, Techniques and Datasets

Suman Kumar Swarnkar and Tien Anh Tran

Abbreviations: remotely operated vehicles (ROVs), autonomous underwater vehicles (AUVs), the peak signal-to-noise ratio (PSNR), the structural similarity index measure (SSIM), discrete wavelet transform (DWT), convolutional neural network (CNN), dark channel prior (DCP), image quality assessment (IQA), ULD Control message (UCM)

1.1 INTRODUCTION

In recent years, academics have focused on underwater exploration. Analysis of underwater images is vital for ocean resource development, marine ecology study, deep-sea installation monitoring, and naval warfare applications [1], [2]. The use of underwater resources is vital to humanity. This has led to the necessity for ROVs and AUVs with high-quality camera systems to investigate the underwater environment [3], [4]. Low-quality underwater images fail image inspection. Developing underwater enhancement methods for advanced underwater imaging is crucial. The last several decades have seen tremendous study in underwater image restoration and enhancement. This chapter reviews underwater image enhancement technologies in detail.

Underwater image processing refines images for human and machine interpretation and processing [5]. Subsea pipeline fracture detection, flora and animal studies, and marine archaeology rely on clear underwater images [6].

Underwater image research depends on the quality of images. It is affected by low visibility, non-uniform illumination, noise, and fading colour [7], [8]. The processing of underwater images is split into two categories: first restoration and second enhancement [1]. For the enhancement category there is no physical model. It uses

DOI: 10.1201/9781003320074-1

subjective criteria to create an appealing image. Enhancement-based methods are easier and quicker [9].

Underwater images must collect crucial data for many uses. Although there are many image enhancement techniques, they are mostly restricted to normal images and relatively few are used for underwater images. Water's environment degrades underwater image quality. With advances in AI and machine learning techniques, underwater image enhancement and restoration have improved significantly [9].

1.2 CHALLENGES OF UNDERWATER IMAGING

Costly underwater investigation requires competent divers. Every inquiry may need standby divers and supervisors. Divers can only spend a certain amount of time underwater. This increases exploring time. This disadvantage may be mitigated using underwater image enhancement.

Water's optical properties hinder underwater photography during light propagation. Refraction makes it harder to concentrate underwater, blurring images. Light and colour absorption alter underwater images. To overcome this constraint, researchers must create effective enhancement strategies.

Water media absorbs a lot of light. Half the light is lost every 10 metres underwater. We have 50% of the light at 10 m and 25% at 20 m [10].

Light availability is time dependent. When the sun is above, water reflects less light. Weather affects light levels. Stormy weather affects light due to uneven water. Light absorption also causes colour absorption.

Absorption reduces image quality by blurring and reducing contrast. High turbidity underwater or intense artificial light sources worsen this difficulty. Artificial light sources provide non-uniform illumination, which causes reflections that conceal image features and generate bright spots [11].

1.3 UNDERWATER IMAGE ENHANCEMENT

Non-physical models manipulate images to improve its enhancement. These algorithms improve underwater images. Underwater image enhancement methods include image colour correction, enhancement and mixed method. The mixed approach combines image colour correction and enhancement method.

1.3.1 Enhancement Methods

Enhancement is a subjective image quality criterion. Two neighbouring planes of differing luminance create contrast. Contrast is the visual difference between an item and its surroundings [12].

Human vision is particularly sensitive to contrast, thus we can perceive in low light. If a picture's contrast is exceptionally strong in one image, such as a black region, information may be lost [13]. Optimising contrast helps show all image details. Numerous techniques for contrast enhancement have been developed for underwater image processing.

[14] suggest a low-frequency sub-band is changed to increase PSNR & SSIM. [15] devised image quality enhancement utilising DWT and DE. Enhancing contrast and brightness begins with contrast enhancement and homomorphic filtering. After DWT decomposition, differential evolution was employed to establish optimum performance evaluation criteria.

CIELAB contains three channels: brightness, chromaticity, and hue. [1] present LAB-MSR, a retinex and LAB-based approach for enhancing underwater picture contrast. MSR combines bilateral and trilateral filter on three CIELAB channels. [5] adjust retinex to eliminate halo artefacts.

[16] model deep CNN networks using convolution layers for encoding and deconvolution layers for decoding [17]. Adaptive improvement ignores environment. Noise-filled photos are handled. [18] decrease distortion, heighten contrast, and increase resolution. MSE, PSNR, and entropy compare well to other methods.

[19] create a low-cost guided filter. Transmission estimate is simplified. [20] use DCP to brighten underwater images. The image's contrast and borders are enhanced.

[21] employ ACCLAHE and DCP to estimate and remove image blur. Homomorphic filtering preserves edges while median filtering removes noise. Regional histogram equalisation solves the problem of real-time enhancement requiring large calculations. The architecture uses a parallel-processing FPGA.

[22] describe a strategy for enhancing each channel's auto levels to increase contrast. This technique is used when decent underwater photography equipment is unavailable. [23] adopt a virtual retina technique for enhancement of image. Contrast enhancement and object colours are obtained. PDCT-based image quality assessment (IQA) is utilised for adaptive enhancement.

[24] improve underwater foggy images by estimating distance. The author achieves significant results by using object distance from a scene point based on pixel intensity and an optimisation strategy.

[25] construct an enhancement model that uses blurriness to calculate the map of depth. The creation model, paired by blurriness data, calculates camera and scene location distance, improving visuals.

[26] enhance colour images using HVS. The method focuses on chromatic components for compressed colour enhancement. Weber-Fechner law is used to remove blocking artefacts. This method differs from others by focusing on chromatic rather than brightness components.

In [27], underwater image enhancement uses particle swarm optimisation to modify RGB. The approach improved underwater image lighting and images. [28] created LIDE based on a parametric model, which reduced memory and processing needs. [28] introduces LIDE with Laplacian mixing model, which improves details and reduces speckle noise (AHE). In [7], a NUCE technique reduces colour cast. A swarm intelligence-based method improves image naturalness. By utilising unsharp masking, the image gets sharpened.

[29] propose a fast enhancement method by using a conditional generative adversarial network (CGAN) based model. They also present image data set EUVP with the collection of poor and good quality images.

A technique based on the properties of light scattering was developed in [30]. First, we divide the colour cast into five groups based on how much of the average of the RGB channels each category represents. Then, in order to repair the colour cast of underwater photographs, we design a multiscene colour restoration approach using the optical attenuation characteristics to determine the colour loss rate of the RGB channels in various situations.

Table 1.1 is a list of the different algorithms that have been made to improve contrast.

TABLE 1.1
Comparison of Different Contrast-based Algorithms for Improvement of Images

Year	Reference	Method	Performance
2012	[31]	EMD	Entropy and gradient values suggest optimal performance.
2012	[26]	HVS	CEF, WBQM, and MASDS values indicate enhancements to colour picture quality.
2014	[32], [33]	Incorporation of picture histogram alterations in RGB and HSV colour models	Acceptable values for entropy, MSE, and PSNR place it second after PDSCC.
2015	[27]	Optimisation using particle swarm	Good performance is shown by PSNR, entropy, and standard deviation values.
2015	[28]	LIDE	A higher level of visual detail and lower speckle noise are created than with AHE.
2016	[34]	Joint guidance image filter	CNR, SSIM, and colour distance measurements demonstrate that this approach effectively eliminates dispersion.
2016	[20]	Enlargement based on DCP	Naturalness image quality evaluator (NIQE) values that are low indicate a high degree of visual acuity.
2016	[21]	ACCLAHE combined with dark channel before	High PSNR values indicate an increase in picture data substance.
2016	[35]	Regional histogram equalisation	Utilisation of resources demonstrates effective installation of ALTIUM nanoboard.
2016	[36]	Differential evolution	Entropy and average gradient values suggest adequate enhancement levels.
2016	[37]	Virtual retina model+ IQA	The results of information entropy, contrast ratio, and PDCT indicate an improvement in contrast and a decent recovery of object colours.
2017	[1]	Encoding-decoding for deep CNN networks	Comparative findings MSE values.
2017	[16]	Interpolation plus modest image improvement	SSIM, PSNR, and MSE values indicate that conv3-deconv3 performs better than conv5-deconv5.
2017	[18]	Image filtering with direction	MSE, PSNR, and entropy levels imply performance excellence.

TABLE 1.1 (Continued)
Comparison of Different Contrast-based Algorithms for Improvement of Images

Year	Reference	Method	Performance
2017	[19]	Transform of stationary wavelet	The design employs 11.3% of gate counts and 42.2% of on-chip memory.
2018	[14]	Wavelet transform algorithm and differential evolution (DEA)	Enhanced PSNR and SSIM.
2019	[15]	Rayleigh elongating	Entropy and PSNR-based techniques have shown to be effective.
2019	[38]	NUCE	MSE and SSIM parameters provide the best outcomes.
2019	[7]	EMD	Entropy, mean gradient, sobel count, and PCQI have the greatest average score, indicating that contrast has been enhanced.
2020	[29]	CGAN	Utilise adversarial training to enhance the quality of images by using information about their global content, colours, local textures, and styles.
2022	[30]	Method based on light scattering characteristics	Categories according to the RGB channels. Then, optical attenuation characteristics to calculate the colour loss rate of the RGB channels of images in different scenes.

1.3.2 Colour Correction Techniques

According to [23], a deep underwater image looks green. Histograms of the three channels show that the red, green and blue. RGB histogram distribution range is not [0,255]. To reduce colour cast in underwater images, a colour correcting procedure must be used.

[39] compare automatic and manual colour correction for underwater pictures. Manual correction is superior to auto-enhancement. Adaptive technique improves picture quality. The proposal outperforms UCM and grey world. Evaluations show bacterial foraging improves visual colour and contrast.

[40] uses non-uniform illumination adjustment. Maximum likelihood estimate maps the picture to Rayleigh distribution. Images contain three channels that is RGB. On three channels, histogram stretching is combined for a corrected picture. AC, AIE, NNF, and CAF are better picture quality metrics than others.

The adaptive linear stretch technique is presented by [41]. Low computational complexity improves subjective quality with this strategy. Low-light zones are corrected using histogram-based thresholds.

[23] created a retinex model for adjusting lighting by extracting a map and applying gamma correction. Adopting a particular normalising procedure corrects image colour cast. When adjusting light, edges are kept, and textures are smoothed. Experiments show that the procedure is better in colour and look than current methods.

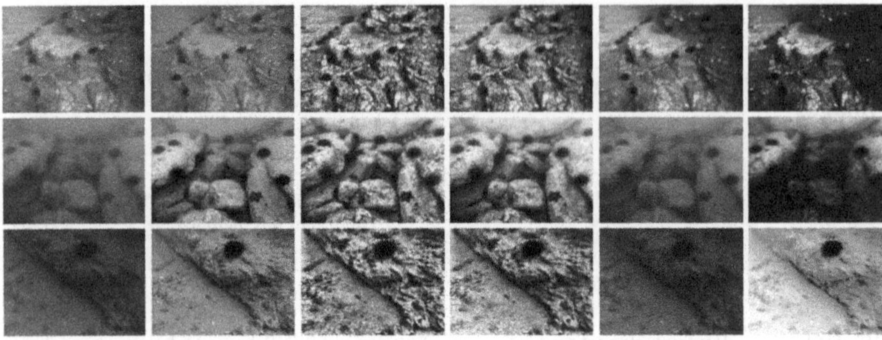

FIGURE 1.1 Colour correction and adaptive contrast enhancement algorithm [43].

[42] adjust colour using wavelets. Approximation and detailed coefficients are derived via DWT on the input image. Colour correction is done with approximation coefficients, and structure preservation is done with detailed coefficients. PSNR and SSIM measurements are high.

Liu et al. use a deep residual framework to improve underwater images [10]. They use CycleGAN and VDSR to improve. Enhancement uses an underwater Resnet model. The approach indicates that batch normalisation (BN) layers increase convergence and enhancement. The highest UIQM score reflects visual effects enhancement.

[43] propose colour correction and adaptive contrast enhancement algorithm. Consider the ratio of the discrepancies between the upper and lower colour channels to the lower colour channels while designing the specialised fractions to compensate the lower colour channels. The resulting image's contrast is then greatly improved by combining the background-and foreground-stretched pictures, which are created by using an adaptive contrast enhancement method to each colour channel. The next step is to apply an unsharp masking method to sharpen the entire image. Shown in Figure 1.1.

Table 1.2 summarises study methods and performance assessment metrics.

1.3.3 Hybrid Methods

Hybrid approaches combine the above-mentioned techniques, such as image enhancement with colour correction or dehazing and colour improvement. Hybrid techniques improve image enhancement. The techniques for enhancing contrast and colour. The contrast/colour adjustment [46] change wavelet coefficients for underwater picture colour correction and contrast enhancement. CCF is generated for each RGB component to modify pixel values. Subjectively and numerically testing the recommended approach using three datasets.

[47] suggested RAHIM for colour and contrast correction. Rayleigh distribution in HSV improves underwater picture contrast. [48] present a two-stage improvement strategy for HSV-transformed images. Rayleigh stretching improves picture details and reduces noise.

TABLE 1.2
Colour Correction Techniques

Year	Reference	Method	Performance
2015	[44]	Fuzzy logic	Outperforms UCM and grey world.
2015	[42]	DWT	PSNR and SSIM are high-quality measures.
2016	[40]	Highest-likelihood projection	AC, AIE, NNF, and CAF show improvement.
2017	[23]	Technique of normalisation	Preserved edges and textures.
2018	[41]	Method of adaptive linear stretching	Entropy, MSE, PSNR,SSIM, and PT represent real-time findings.
2019	[45]	Cycle VDSR+ GAN	High UIQM means visual enhancement.
2021	[43]	Colour correction and adaptive contrast enhancement algorithm	Produce the background-stretched and foreground.

[49] offer a residual CNN for transmission map estimation. URCNN helps balance light. Image haze is removed and contrast is increased. [50] present a single-image approach for underwater Rayleigh stretching. Underwater colour cast is corrected using white balance and gamma. Overexposed and underexposed picture portions are minimised, according to qualitative and quantitative analyses.

[51] employ multiscale fusion to improve global contrast and edge sharpness. It works great without camera adjustments. [52] use empirical mode decomposition to minimise scattering and blurring caused to inadequate light. A white balancing with grey world method is used to address contrast and colour cast difficulties.

[53] enhance underwater photos without magnifying artefacts. The recommended approach uses histogram equalisation for colour correction and TV-L1 for image decomposition into structure and decomposition layers. Texture mask is created by denoising and improving these layers. [54] correct colour distortion using PLT. [53] reduce artefacts and improve the underwater picture with optimal contrast. Real-time computation is implied.

Frequency fusion enhances underwater photos [23]. Fusing uses wavelet decomposition, weighted average, and local variance. [55] present a frequency-domain approach for colour and contrast correction. DCT boosts visual contrast. UIQM data show the system extracts key undersea elements for maritime border security.

[56] create a contrast-enhanced, white-balanced picture from a hazy underwater shot. Two input pictures are weighted by chrominance, saliency, and brightness. The final dehazed picture is created by fusing input photos and weight maps. [57] employ hybrid DCT-DHE. Before block splitting, a dynamic histogram equalisation filter is used.

[58] decomposed colour-corrected, contrast-enhanced pictures to derive wavelet coefficients. The method fuses maximal coefficients from deconstructed pictures. RMSE, PSNR, and SSIM improve with inverse decomposition.

Nonlinear filtering improves muddy water photos. Automation corrects saturation, lighting, contrast, and colour. AMBE and EME reveal that the procedure is more effective than prior ones. Using a fractional filter, [59] boost contrast. Strong and

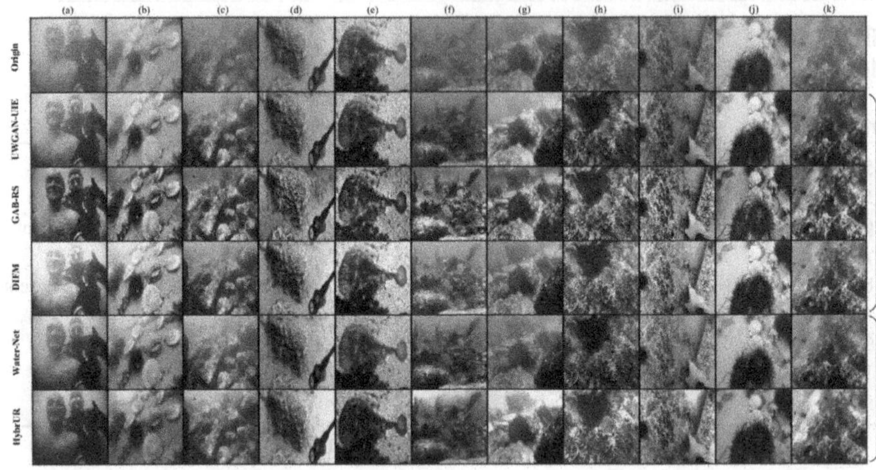

FIGURE 1.2 Underwater image enhancement using hybrid method [63].

weak edges affect filter order. Adaptive fractional differential function improves weak textures and edges.

[60] increase deep underwater images' colour cast and contrast using RCCGB and mean pixel enhancement. PSO enhances contrast while RCCGB lowers lighting effects. NIQE, entropy, and average gradient indicate strong performance.

[61] use a CGAN to enhance multiscale photos. Dual discriminators acquire local and global information and naturalise images. PSNR and SSIM outperform other current methods.

The goal of [62]'s system is to better correct for attenuation differences throughout the propagation channel and eliminate unwanted colour casts by integrating the enhanced underwater white balance (UWB) method with histogram stretching.

[63] provide an unsupervised architecture that learns underwater vision restoration from unpaired underwater-terrestrial pictures. The underwater degeneration must be specifically built based on the optically clear physics law in order to sufficiently convert the domain and preserve detail. Shown in Figure 1.2.

Table 1.3 shows hybrid approaches for enhancing contrast and colour. The table shows the techniques' pros and cons.

1.4 UNDERWATER IMAGE DATASETS

MBARI dataset [2] contains fish images. That is available on www.mbari.org. And, the RUIE data set, with over 4,000 images [45] has 3 subsets. Image visibility, quality, colour casts, and high-level detection and classification are addressed.

AFSC dataset collected by an underwater ROV using an RGB camera. This dataset contains several ROV footage. AFSC has 571 images related to fish [2]. Images have a 2112 x 2816 resolution.

TABLE 1.3
Colour and Contrast Correction Methods

Year	Reference	Method	Performance
2012	[8]	Multiscale fusing	The IQM metric amplifies contrast.
2014	[32]	Image histogram is changed to RGB and HSV	Entropy, MSE, and PSNR represent picture value.
2015	[33]	DIRSCLAHS dual-image Adaptive Rayleigh-stretched contrast histogram	High EME and entropy suggest picture detail enhancement.
2016	[34]	Guided trigonometric weights	PSNR and SSIM demonstrate the approach reduces dispersion.
2017	[46]	Wavelet-based variant improvement	PSNR, entropy, and SSIM indexes reveal object contrast and structure.
2017	[47]	Adaptive histogram mod (RAHIM)	EME, EMEE, NIQE, and entropy have the finest visual quality.
2017	[52]	Decomposition of modes (EMD)	Low MSE and high PSNR enhance vision.
2018	[49]	URCNN	Image quality is shown by low blur metric (BM), SSEQ, and NIQE scores.
2018	[64]	CLAHE+ percentile technique	Good outcomes are shown by RMSE and entropy.
2018	[50]	Combining white balance with light scattering stretching	NIQE and UICM readings are excellent.
2018	[50]	Single-image method without dedicated hardware	PCQI, UCIQE, and UIQM suggest strong perceptual quality, global contrast improvement, and image structural detail.
2019	[65]	Blending	MSE, PSNR, and entropy indicate a decent fused picture.
2019	[60]	RCCGB+SCSMPE	Visual improvement is shown by NIQE, entropy, and MSSIM.
2020	[62]	UWB+VCSE	Integrated with histogram stretching.
2021	[63]	Data and physics driven unsupervised method	Overcome the problem of invalid gradient when optimising the hybrid physical-neural model.

Remotely controlled vehicles stare at the ocean bottom to gather **NWFSC** data. Initial data release includes 123 images of fish and other creatures along the seafloor. The collection contains 2448 by 2050 images [2]. Instead of bounding boxes, this dataset has important points.

The **MOUSS** data that [2] analysed was captured by a stationary camera that was positioned 1–2 metres above the sea bottom. The collection contains 159 labelled fish images. Test data without comments contain training and new images. Images are 968 x 728.

CVPR AAMVEM integrates AFSC, NWFSC, and MOUSS databases. The SUN database contains annotated images of underwater settings, locations, and items [66].

[1] utilised internet-sourced raw underwater images. 45 deteriorated underwater images. [28] provides a link to the whole dataset.

A real undersea dataset from the University of Michigan was used by [67]. More than 15,000 underwater photos. The **SQUID** dataset, which includes 57 stereo pairs from four sites in Israel, includes two stereo pairs from the Red Sea, which represents tropical water, and two stereo pairs from the Mediterranean, which represents temperate water. A shipwreck and coral reef are in the Red Sea. In the Mediterranean Sea, Nachsholim is 3–6 metres deep while Mikhmoret is 10–12 metres deep [68].

950 underwater photos comprise the Underwater Image Enhancement Benchmark (**UIEB**) in Li et al. [62]. Reference labels are included on 890 underwater pictures. Since there were no accessible reference photographs, the final 60 undersea images are challenging. To analyse underwater picture enhancing techniques, utilise this dataset.

Unpaired and paired underwater pictures are included in the Enhancement of Underwater Visual Perception (**EUVP**). These pictures are taken by seven cameras during deep-sea research and human-robot experiments. The dataset supports training against the system. The dataset may be found at http://irvlab.cs.umn.edu/resources/euvp-dataset.

The **TURBID** [69] dataset was introduced for underwater image processing research. It includes five partitions of distorted images and ground-truth. Knowledge contains live video fish data. It includes 27,370 confirmed fish images and is divided into 23 clusters, each representing a species [70]. Table 1.4 shows an example of data.

TABLE 1.4
Datasets Used by Researchers for Experiment

SN	Dataset	Full Name	Object in Dataset	Source	No of Images
1.	MBARI	Monterey Bay Aquarium Research Institute	Fishes	mbari.org	666
2.	RUIE	Real-world Underwater Image Enhancement	Different underwater images	-	4000
3.	MOUSS	Modular Optical *Underwater* Survey System	Fishes	CVPR AAMVEM workshop	159
4.	NWFSC	Northwest Fisheries Science Center	Fishes, other related species	www.fisheries.noaa.gov/inport/stats/NWFSC/data-sets	123
5.	AFSC	Alaska Fisheries Science Center	Fishes, other related species	www.fisheries.noaa.gov/inport/stats/AFSC/data-sets	571
6.	UIEB	Underwater Image Enhancement Benchmark	Diverse species	https://li-chongyi.github.io/proj_benchmark.html	950
7.	TURBID	TURBID	Diverse species	http://amandaduarte.com.br/turbid/	666

1.5 CONCLUSION

This study examines underwater image enhancement efforts. We explore the underwater imaging model and how fading colours degrade underwater images. The survey classifies and presents the researchers' enhancement strategies. This chapter describes the datasets accessible for experimentation and study. Image quality criteria needed to evaluate underwater image quality are presented and calculated. Two datasets were evaluated using state-of-the-art approaches to explain image enhancement. This chapter discusses recent implementations of underwater image enhancement methods to raise awareness of the need for underwater image processing and aid future researchers.

In recent years, algorithms for single underwater image enhancement approaches have grown rapidly, but no algorithm has been devised to enhance images collected in diverse underwater habitats and depths. Traditional underwater enhancement methods lack flexibility and resilience. Reduce the complexity of the procedures to scale up work for practical research and applications. Combining image enhancement and restoration improves enhancement techniques.

ACKNOWLEDGEMENTS

The authors acknowledge the editor of this book for a critical reading of the manuscript and the reviewers for their critical observations. I also gratefully acknowledge the co-author for his contribution.

REFERENCES

[1] H. Lu, Y. Li, Y. Zhang, M. Chen, S. Serikawa, and H. Kim, "Underwater optical image processing: A comprehensive review," *Mob. Networks Appl.*, vol. 22, no. 6, 2017, doi: 10.1007/s11036-017-0863-4

[2] M. Yang, J. Hu, C. Li, G. Rohde, Y. Du, and K. Hu, "An in-depth survey of underwater image enhancement and restoration," *IEEE Access*, vol. 7, 2019, doi: 10.1109/ACCESS.2019.2932611

[3] J. Ahn, S. Yasukawa, T. Sonoda, T. Ura, and K. Ishii, "Enhancement of deep-sea floor images obtained by an underwater vehicle and its evaluation by crab recognition," *J. Mar. Sci. Technol.*, vol. 22, no. 4, pp. 758–770, Dec. 2017, doi: 10.1007/s00773-017-0442-1

[4] G. Johnsen, M. Ludvigsen, A. Sørensen, and L. M. Sandvik Aas, "The use of underwater hyperspectral imaging deployed on remotely operated vehicles–methods and applications," in *IFAC-PapersOnLine*, vol. 49, no. 23, 2016, doi: 10.1016/j.ifacol.2016.10.451

[5] W. Zhang, L. Dong, X. Pan, P. Zou, L. Qin, and W. Xu, "A survey of restoration and enhancement for underwater images," *IEEE Access*, vol. 7, 2019, doi: 10.1109/ACCESS.2019.2959560

[6] A. Khan, S. S. A. Ali, A. Anwer, S. H. Adil, and F. Meriaudeau, "Subsea pipeline corrosion estimation by restoring and enhancing degraded underwater images," *IEEE Access*, vol. 6, 2018, doi: 10.1109/ACCESS.2018.2855725

[7] K. Z. Mohd Azmi, A. S. Abdul Ghani, Z. Md Yusof, and Z. Ibrahim, "Natural-based underwater image color enhancement through fusion of swarm-intelligence algorithm," *Appl. Soft Comput. J.*, vol. 85, 2019, doi: 10.1016/j.asoc.2019.105810
[8] C. Ancuti, C. O. Ancuti, T. Haber, and P. Bekaert, "Enhancing underwater images and videos by fusion," 2012, doi: 10.1109/CVPR.2012.6247661
[9] S. Anwar and C. Li, "Diving deeper into underwater image enhancement: A survey," *Signal Process. Image Commun.*, vol. 89, 2020, doi: 10.1016/j.image.2020.115978
[10] P. Liu, G. Wang, H. Qi, C. Zhang, H. Zheng, and Z. Yu, "Underwater image enhancement with a deep residual framework," *IEEE Access*, vol. 7, 2019, doi: 10.1109/ACCESS.2019.2928976
[11] S. Corchs and R. Schettini, "Underwater image processing: State of the art of restoration and image enhancement methods," *EURASIP J. Adv. Signal Process.*, vol. 2010, 2010, doi: 10.1155/2010/746052
[12] Y. Ueki and M. Ikehara, "Underwater image enhancement based on the iteration of a generalization of dark channel prior," 2019, doi: 10.1109/VCIP47243.2019.8965726
[13] H. Liu and L. P. Chau, "Underwater image restoration based on contrast enhancement," in *International Conference on Digital Signal Processing, DSP*, 2016, doi: 10.1109/ICDSP.2016.7868625
[14] R. Priyadharsini, T. Sree Sharmila, and V. Rajendran, "A wavelet transform based contrast enhancement method for underwater acoustic images," *Multidimens. Syst. Signal Process.*, vol. 29, no. 4, 2018, doi: 10.1007/s11045-017-0533-5
[15] G. E. Guraksin, O. Deperlioglu, and U. Kose, "A novel underwater image enhancement approach with wavelet transform supported by differential evolution algorithm," in *Intelligent Systems Reference Library*, vol. 150, 2019
[16] X. Sun, L. Liu, and J. Dong, "Underwater image enhancement with encoding-decoding deep CNN networks," in *2017 IEEE SmartWorld, Ubiquitous Intelligence & Computing, Advanced & Trusted Computed, Scalable Computing & Communications, Cloud & Big Data Computing, Internet of People and Smart City Innovation* (SmartWorld/SCALCOM/UIC/ATC/CBDCom/IOP/SCI), Aug. 2017, pp. 1–6, doi: 10.1109/UIC-ATC.2017.8397462
[17] E. Goceri, "Skin disease diagnosis from photographs using deep learning," in *Lecture Notes in Computational Vision and Biomechanics*, vol. 34, 2019.
[18] A. Bindhu and O. U. Maheswari, "Under water image enhancement based on linear image interpolation and limited image enhancer techniques," 2017, doi: 10.1109/ICSCN.2017.8085692
[19] C. B. H. Chang, I. C. Huang, M. F. Lin, and S. R. Kuang, "Design and implementation of a low-cost guided image filter for underwater image enhancement," 2017, doi: 10.1109/DESEC.2017.8073850
[20] X. Li, Z. Yang, M. Shang, and J. Hao, "Underwater image enhancement via dark channel prior and luminance adjustment," 2016, doi: 10.1109/OCEANSAP.2016.7485625
[21] S. Dixit, S. K. Tiwari, and P. Sharma, "Underwater image enhancement using DCP with ACCLAHE and homomorphism filtering," 2017, doi: 10.1109/SCOPES.2016.7955807
[22] A. Basuki and N. Ramadijanti, "Improving auto level method for enhancement of underwater images," 2017, doi: 10.1109/KCIC.2016.7883635
[23] S. Zhang, T. Wang, J. Dong, and H. Yu, "Underwater image enhancement via extended multi-scale Retinex," *Neurocomputing*, vol. 245, 2017, doi: 10.1016/j.neucom.2017.03.029
[24] P. K. Dwivedi, B. Paul, and D. Ghoshal, "Notice of removal: Underwater image enhancement using distance factor estimation," *International Conference on*

Electrical, Electronics, Signals, Communication and Optimization, EESCO 2015. 2015, doi: 10.1109/EESCO.2015.7254047
[25] Y. T. Peng, X. Zhao, and P. C. Cosman, "Single underwater image enhancement using depth estimation based on blurriness," in *Proceedings - International Conference on Image Processing, ICIP*, 2015, vol. 2015-December, doi: 10.1109/ICIP.2015.7351749
[26] O. C. Au, L. Sun, R. Zou, W. Dai, and S. Li, "An improved method for color images enhancement considering HVS," 2012, doi: 10.1109/ICALIP.2012.6376597
[27] A. Abunaser, I. A. Doush, N. Mansour, and S. Alshattnawi, "Underwater image enhancement using particle swarm optimization," *J. Intell. Syst.*, vol. 24, no. 1, 2015, doi: 10.1515/jisys-2014-0012
[28] S. Marukatat, "Image enhancement using local intensity distribution equalization," *Eurasip J. Image Video Process.*, vol. 2015, no. 1, 2015, doi: 10.1186/s13640-015-0085-2
[29] M. J. Islam, Y. Xia, and J. Sattar, "Fast underwater image enhancement for improved visual perception," *IEEE Robot. Autom. Lett.*, vol. 5, no. 2, 2020, doi: 10.1109/LRA.2020.2974710
[30] J. Zhou, X. Wei, J. Shi, W. Chu, and W. Zhang, "Underwater image enhancement method with light scattering characteristics," *Comput. Electr. Eng.*, vol. 100, 2022, doi: 10.1016/j.compeleceng.2022.107898
[31] A. T. Çelebi and S. Ertürk, "Visual enhancement of underwater images using Empirical Mode Decomposition," *Expert Syst. Appl.*, vol. 39, no. 1, 2012, doi: 10.1016/j.eswa.2011.07.077
[32] A. S. A. Ghani and N. A. M. Isa, "Underwater image quality enhancement through Rayleigh-stretching and averaging image planes," *Int. J. Nav. Archit. Ocean Eng.*, vol. 6, no. 4, pp. 840–866, Dec. 2014, doi: 10.2478/IJNAOE-2013-0217
[33] A. S. Abdul Ghani and N. A. Mat Isa, "Underwater image quality enhancement through integrated color model with Rayleigh distribution," *Appl. Soft Comput. J.*, vol. 27, 2015, doi: 10.1016/j.asoc.2014.11.020
[34] H. Lu et al., "Underwater image enhancement method using weighted guided trigonometric filtering and artificial light correction," *J. Vis. Commun. Image Represent.*, vol. 38, pp. 504–516, Jul. 2016, doi: 10.1016/j.jvcir.2016.03.029
[35] R. S. M. Alex, S. Deepa, and M. H. Supriya, "Underwater image enhancement using CLAHE in a reconfigurable platform," 2016, doi: 10.1109/OCEANS.2016.7761194
[36] G. E. Guraksin, U. Kose, and O. Deperlioglu, "Underwater image enhancement based on contrast adjustment via differential evolution algorithm," 2016, doi: 10.1109/INISTA.2016.7571849
[37] Y. Wang et al., "Underwater image enhancement strategy with virtual retina model and image quality assessment," 2016, doi: 10.1109/OCEANS.2016.7761381
[38] S. Sankpal and S. Deshpande, "Underwater image enhancement by Rayleigh stretching with adaptive scale parameter and energy correction," in *Adv. Intell. Syst. Comput.*, vol. 810, 2018.
[39] N. Bt Shamsuddin, W. F. Bt Wan Ahmad, B. B. Baharudin, M. K. B. M. Rajuddin, and F. Bt Mohd, "Significance level of image enhancement techniques for underwater images," in *2012 International Conference on Computer and Information Science, ICCIS 2012 - A Conference of World Engineering, Science and Technology Congress, ESTCON 2012 - Conference Proceedings*, 2012, vol. 1, doi: 10.1109/ICCISci.2012.6297295
[40] S. S. Sankpal and S. S. Deshpande, "Nonuniform illumination correction algorithm for underwater images using maximum likelihood estimation method," *J. Eng. (United Kingdom)*, 2016, doi: 10.1155/2016/5718297

[41] J. Ao and C. Ma, "Adaptive stretching method for underwater image color correction," *Int. J. Pattern Recognit. Artif. Intell.*, vol. 32, no. 2, 2018, doi: 10.1142/S0218001418540010
[42] G. Singh, N. Jaggi, S. Vasamsetti, H. K. Sardana, S. Kumar, and N. Mittal, "Underwater image/video enhancement using wavelet based color correction (WBCC) method," 2015, doi: 10.1109/UT.2015.7108303
[43] W. Zhang, X. Pan, X. Xie, L. Li, Z. Wang, and C. Han, "Color correction and adaptive contrast enhancement for underwater image enhancement," *Comput. Electr. Eng.*, vol. 91, p. 106981, May 2021, doi: 10.1016/j.compeleceng.2021.106981
[44] R. Sethi, I. Sreedevi, O. P. Verma, and V. Jain, "An optimal underwater image enhancement based on fuzzy gray world algorithm and Bacterial Foraging algorithm," 2016, doi: 10.1109/NCVPRIPG.2015.7490004
[45] R. Liu, X. Fan, M. Zhu, M. Hou, and Z. Luo, "Real-world underwater enhancement: Challenges, benchmarks, and solutions under natural light," *IEEE Trans. Circuits Syst. Video Technol.*, vol. 30, no. 12, 2020, doi: 10.1109/TCSVT.2019.2963772
[46] S. Vasamsetti, N. Mittal, B. C. Neelapu, and H. K. Sardana, "Wavelet based perspective on variational enhancement technique for underwater imagery," *Ocean Eng.*, vol. 141, 2017, doi: 10.1016/j.oceaneng.2017.06.012
[47] A. S. Abdul Ghani and N. A. Mat Isa, "Automatic system for improving underwater image contrast and color through recursive adaptive histogram modification," *Comput. Electron. Agric.*, vol. 141, 2017, doi: 10.1016/j.compag.2017.07.021
[48] A. S. A. Ghani and N. A. M. Isa, "Underwater image quality enhancement through Rayleigh-stretching and averaging image planes," *Int. J. Nav. Archit. Ocean Eng.*, vol. 6, no. 4, pp. 840–866, Dec. 2014, doi: 10.2478/IJNAOE-2013-0217
[49] M. Hou, R. Liu, X. Fan, and Z. Luo, "Joint residual learning for underwater image enhancement," 2018, doi: 10.1109/ICIP.2018.8451209
[50] M. Mathur and N. Goel, "Enhancement of underwater images using white balancing and Rayleigh-stretching," 2018, doi: 10.1109/SPIN.2018.8474042
[51] C. O. Ancuti, C. Ancuti, C. De Vleeschouwer, and P. Bekaert, "Color balance and fusion for underwater image enhancement," *IEEE Trans. Image Process.*, vol. 27, no. 1, 2018, doi: 10.1109/TIP.2017.2759252
[52] S. Mallik, S. S. Khan, and U. C. Pati, "Visual enhancement of underwater image by white-balanced EMD," 2017, doi: 10.1109/ICCCNT.2017.8204163
[53] J. Yang, X. Wang, H. Yue, X. Fu, and C. Hou, "Underwater image enhancement based on structure-texture decomposition," in *Proceedings–International Conference on Image Processing, ICIP*, vol. 2017-September, 2018doi: 10.1109/ICIP.2017.8296473
[54] X. Fu, Z. Fan, M. Ling, Y. Huang, and X. Ding, "Two-step approach for single underwater image enhancement," in *2017 International Symposium on Intelligent Signal Processing and Communication Systems, ISPACS 2017–Proceedings*, vol. 2018-January, 2017, doi: 10.1109/ISPACS.2017.8266583
[55] O. Erat, K. Panetta, and S. Agaian, "Contrast enhancement for underwater images in maritime border protection," in *2017 IEEE International Symposium on Technologies for Homeland Security, HST 2017*, vol. 2017, April 2017, doi: 10.1109/THS.2017.8050197
[56] R. Singh and M. Biswas, "Contrast and color improvement based haze removal of underwater images using fusion technique," in *4th IEEE International Conference on Signal Processing, Computing and Control, ISPCC 2017*, vol. 2017, January, 2017, doi: 10.1109/ISPCC.2017.8269664
[57] N. Dubey, S. K. Tiwari, and P. Sharma, "A hybrid DCT-DHE approach for enhancement of low contrast underwater images," in *International Conference on Recent*

Innovations in Signal Processing and Embedded Systems, RISE 2017 , vol. 2018, January, 2018, doi: 10.1109/RISE.2017.8378171
[58] A. Khan, S. S. A. Ali, A. S. Malik, A. Anwer, and F. Meriaudeau, "Underwater image enhancement by wavelet based fusion," 2017, doi: 10.1109/USYS.2016.7893927
[59] S. Shourya, S. Kumar, and R. K. Jha, "Adaptive fractional differential approach to enhance underwater images," 2017, doi: 10.1109/ISED.2016.7977054
[60] K. Z. Mohd Azmi, A. S. Abdul Ghani, Z. Md Yusof, and Z. Ibrahim, "Deep underwater image enhancement through colour cast removal and optimization algorithm," *Imaging Sci. J.*, vol. 67, no. 6, 2019, doi: 10.1080/13682199.2019.1660484
[61] M. Yang, K. Hu, Y. Du, Z. Wei, Z. Sheng, and J. Hu, "Underwater image enhancement based on conditional generative adversarial network," *Signal Process. Image Commun.*, vol. 81, p. 115723, Feb. 2020, doi: 10.1016/j.image.2019.115723
[62] X. Li, G. Hou, L. Tan, and W. Liu, "A hybrid framework for underwater image enhancement," *IEEE Access*, vol. 8, pp. 197448–197462, 2020, doi: 10.1109/ACCESS.2020.3034275
[63] S. Yan et al., "HybrUR: A Hybrid Physical-Neural Solution for Unsupervised Underwater Image Restoration," Jul. 2021, Accessed: Jul. 26, 2022. [Online]. Available: http://arxiv.org/abs/2107.02660.
[64] D. Garg, N. K. Garg, and M. Kumar, "Underwater image enhancement using blending of CLAHE and percentile methodologies," *Multimed. Tools Appl.*, vol. 77, no. 20, 2018, doi: 10.1007/s11042-018-5878-8
[65] P. Honnutagi, V. D. Mytri, and Y. S. Lalitha, "Fusion-based underwater image enhancement by weight map techniques," in *Adv. Intell. Syst. Comput.*, vol. 740, 2019.
[66] J. Xiao, J. Hays, K. A. Ehinger, A. Oliva, and A. Torralba, "SUN database: Large-scale scene recognition from abbey to zoo," 2010, doi: 10.1109/CVPR.2010.5539970
[67] C. Li, J. Guo, C. Guo, R. Cong, and J. Gong, "A hybrid method for underwater image correction," *Pattern Recognit. Lett.*, vol. 94, 2017, doi: 10.1016/j.patrec.2017.05.023
[68] D. Berman, D. Levy, S. Avidan, and T. Treibitz, "Underwater single image color restoration using haze-lines and a new quantitative dataset," *IEEE Trans. Pattern Anal. Mach. Intell.*, vol. 43, no. 8, 2021, doi: 10.1109/TPAMI.2020.2977624
[69] A. Duarte, F. Codevilla, J. D. O. Gaya, and S. S. C. Botelho, "A dataset to evaluate underwater image restoration methods," 2016, doi: 10.1109/OCEANSAP.2016.7485524
[70] B. J. Boom et al., "Long-term underwater camera surveillance for monitoring and analysis of fish populations," Work. Vis. Obs. Anal. Anim. Insect Behav. (VAIB), conjunction with ICPR 2012, 2012.

2 Advances of Unmanned Autonomous Vehicle and Its Communication Systems

B. Akoramurthy and K. Dhivya

2.1 INTRODUCTION

Unmanned Aerial Vehicles (UAV), are air shuttles with no human pilot aboard, which are completely controlled and successfully operated remotely by the software. UAVs can also be called by other names, such as drones, zeppelins, blimps, dirigibles, etc. Drones are a contemporary archetype that emerged from the field of in-flight robotics and have a great deal of promise to open up newer applications and economic prospects [1]. The global aerial market, which was valued at around $20 billion in 2018, is expected to grow to $60 billion by the next decade at a mean annual growth rate of nearly 14 per cent over a set period of years [2]. Highly desirable drones, because of their easiness of deployment and mobility, have been used in a variety of use cases, such as border security, counter insurgency in countries' borders, aerial policing for crime control and prevention, surveillance, creation of adhoc networks in telecommunication, search and salvage missions (SAS), and so on. The majority of UAS courses begin with a cataloguing of drones using a type of traditional categorisation, as follows: Tactical or military grade drones are drones based on flight height and durability, and also based upon their size. The primary disadvantage of such metaphors is that they mostly draw from already-existing systems, combining size (small or micro), altitude, and mission capabilities (VTOL, long endurance). Quite a classification will not provide a detailed overview of the various options available when it comes to operations and vehicle configurations. Furthermore, the life of drones is based on the drone market dissection that is currently in place, which makes it exceptionally grim to predict the future of drones. An input matrix that combines archetypal job patterns and drone arrangements would be a more acceptable method to categorise the various types of potential UAS.

 DOI: 10.1201/9781003320074-2

Mission profiles may include the following:

- Find out about missions (both outdoor and indoor) that necessitate VTOL capabilities.
- Surveillance missions (including any range) that necessitate long endurance,
- Other specific missions include shipping supplies, tracking special facilities ranging from windmills to nuclear power plants, and some tactical military missions mandating covertness (low noise and detection and ranging signature) and robust transmission.

It should be noted that when it comes to mission profiles, the majority of end users find it challenging to really define their mission needs without also turning to a previous specification of a configuration. Appropriately differentiating between the payload/vehicle description and the mission criteria is crucial during the UAS design phase. For instance, it is necessary to clearly state the dimensions of the space utilised, the range from the take-off zone to the region of interest, and the optimum period of time allowed to gather the needed data, in addition to supplemental potential limitations pertaining to logistics, regulations, overheads, etc. when surveying a secluded spot in the ocean. The choice of selection needs to be based upon the location of interest. A most efficient drone must be selected if the remote location is far from the landing area. The network may be composed of a wide variety of vehicles (which might be smaller or larger in terms of performance and ability), each of which has a finite lifetime but can give almost infinite eavesdropping capabilities by switching off between them. If the location chosen is not a distant one, then constant monitoring will be needed to make the system more effective. The latter solution might offer a significantly better cost-performance trade-off than the previous one. In actuality, a small aircraft, which is easier to mobilise than a larger one, might be supplied with a less affordable burden since it is concentrated on a significantly constrained observational site. The three primary kinds of vehicle configurations are defined in the following section, along with the hybrid configuration of drones. The majority of the vehicles in the fourth category will be convertibles with tilt-rotor, tilt-wing, or tilt-body platforms. The majority of extant ornithopters will also be included, which typically include fluttering arms and a permanent exterior that serves as a tail or elevator. Other aircraft designs, such as blimps and skydivers, will be considered as a discrete class even if they make up a smaller portion of current and future drone systems.

2.1.1 Permanent-arms Drones

Fixed-wing UAVs range from micro-sized drones (Figure 2.1), also known as micro-air vehicles (MAVs), to UAVs that are nearly as large as any current conventional aircraft. The Wasp, developed by AeroVironment, is a **forty-one-centimeter-long** chargeable powered flying arm of two hundred and seventy-five grams and serves as an illustration of a small permanent arm drone (Figure 2.2). Even more compact fixed-wing MAVs were created, such as the flexible-wing MAV with an X-cm span from the Florida university around 2004–05.

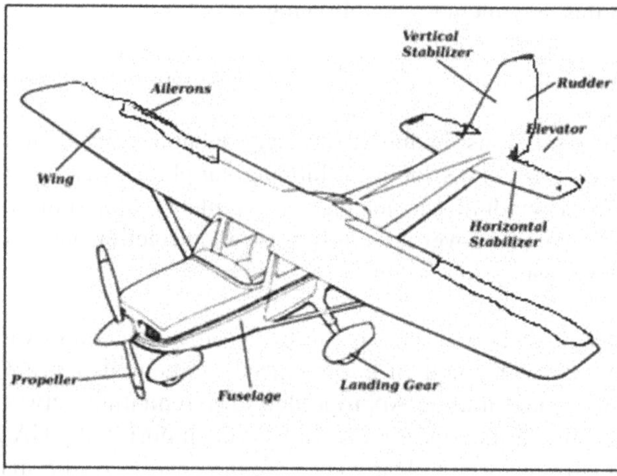

FIGURE 2.1 Micro-sized UAV.

FIGURE 2.2 Compact fixed-wing MAVs–with 10 cm span. (Photo credit: Aselsan.)

The "SolarEagle" (Figure 2.3) was designed to be a "satellite-drone" that could fly for nearly a whole day owing to solar panels covering the upper half of its arms and extremely severe limits on structural design to make it as tiny as desirable. This is in contrast to the incredibly small fixed-wing UAVs.

The 426.509-foot-span fixed-wing solar-powered drone must overcome the well-known mathematical principle, which states that volume grows faster than air foil surface. Based on the noted reason, sun-powered drones will be suitable at lower dimensions as, in comparison to larger aircraft, a greater amount of the power required to fuel the motor may be supplied from the sun. For instance, the Solar-Storm, a 1.64042 foot-span permanent arm coated in tiny, light sun-powered batteries, was created to increase the durability of an existing model that is totally powered by conventional batteries. On bright days, the Solar-Storm (see Figure 2.4) [3] could provide up to nearly 46 per cent of the total electricity required to fly.

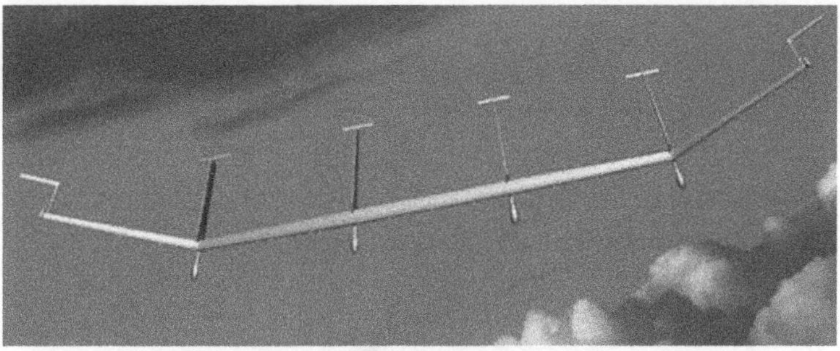

FIGURE 2.3 Solar Eagle. (Photo credit: Boeing.)

FIGURE 2.4 Solar-powered UAV. (Photo credit: Murat Bronx.)

It is important to note, from a practical standpoint, that these tiny solar-powered vehicles don't need a battery charger that must be connected to an electrical source. While one mini-UAV is in the air, another of the same type could be recharging on the ground. In spite of their inherent difficulty in hovering, fixed-wing UAVs are still much better prospects than rotary-wing UAVs for distant observation jobs. Even palm-sized intermediate permanent arm drones (approximately below 10 kg) remain in the air for up to eight hours each day, which is typically sufficient for a standard surveillance mission. Although the engineering process of designing an airplane is already well recognised for orthodox airplanes, it may not be completely understood for smaller drones due to low-slung Re effects (Re) that impair flying style and momentum. It's worth mentioning that in order to get satisfactory results, thorough design and manufacturing techniques must be applied and tailored to the area of drones. Furthermore, for long-endurance needs, substantial values of the proportion, such as some mathematical quotations, are required. Because of this, long-range permanent arm drones are linked to relatively large mathematical values and may cause trip circumstances that are nearly stall-like. As a result, while developing a long-range drone, a capacity/weight factor, lifting off from the ground, and the efficiency of getting back to the

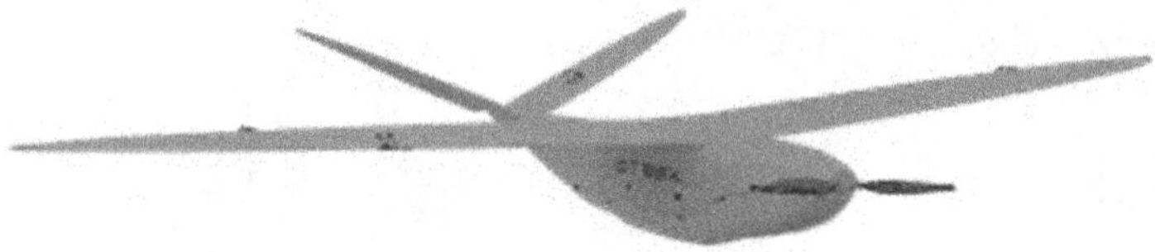

FIGURE 2.5 DT26 Open Payload. (Credit: Delair-Tech.)

FIGURE 2.6 Concept for a long-range mini-UAV was developed after the albatross flying style. (Photo courtesy Philip Richardson, 2012.)

ground should be considered. Then, as shown in Figure 2.5, which depicts the permanent arm drone DT26X open payload drone designed by Delair-Technology [4], detailed wind tunnel testing and enhancement processes will be carried out.

In addition to recent advancements in fuel cell downsizing, extracting energy from the atmosphere is an intriguing strategy for significantly extending mini-UAVs' endurance.

Thermals, like in the case of sailplanes, or shear wind, can be used to capture energy. The albatross flight, which takes advantage of shear wind produced by the upper layer of the atmosphere over the ocean's surface, is the best illustration of such a mechanism in nature. We now have a better understanding of this process, also known as dynamic soaring, and we can mimic it analytically. Some researchers have proposed that drones will be utilised for oceanic observation, investigation, and quest-and-salvage jobs could be made using the concepts of dynamic soaring (see Figure 2.6).

2.1.2 Flapping-wing UAVs

Some authors have proposed that existing flying species, such as birds or insects, should serve as inspiration for engineers since the very beginning of flight. Such a viewpoint is predicated on the concept that animals have gradually improved over time.

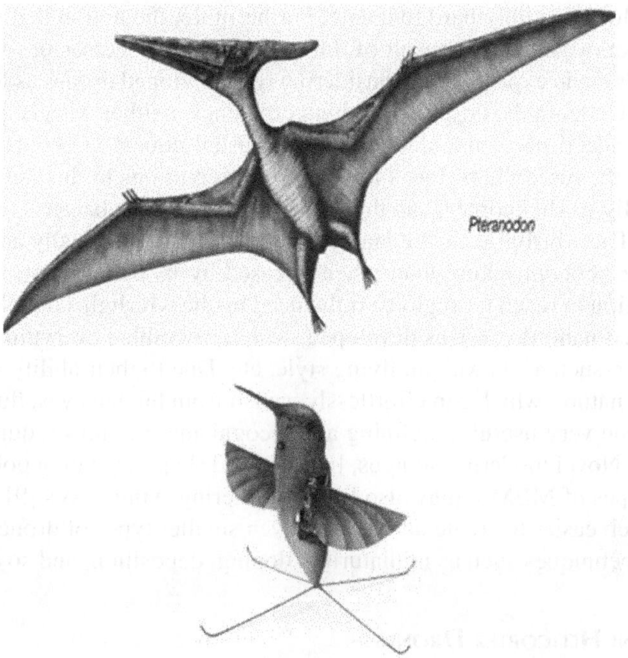

FIGURE 2.7a (Left) Pteranodon-flying dinosaur inspiration for the UAV (Right) Spy Humming bird which is nano UA-Vehicle.

Hundreds of species, including the famed Pteranodon (Figure 2.7a), a winged dino with close to 700 cm wing length, although its exact weight is still debatable, to the fairy flies, the cutest winged bug at only 0.14986 mm [5], provide enthralling specimens of any size of animal. The hummingbird deserves special attention because it served as an inspiration for AeroVironment's recently created nano air vehicle (Figure 2.7b) [6].

The inherent complexity of the flow field and the resulting unsteadiness make it difficult to fully understand the aerodynamics of flapping wings. It has been the focus of several researchers or academicians for the last four decades, including a range of experimental and numerical methodologies [7].

Despite the fact that current bugs and birds do not hover [6] as efficiently as typical rotors, even at their relatively lowest Re, it is not clear that fluttering flight is truly more economical than rotary-wing systems. Furthermore, a new study indicates that some bugs' fluttering flight may be substantially less successful than previously thought [8]. The beginning and ending places of the fluttering wave have very little flying competence due to the wind movement relative to the aircraft decreasing at that location, which may be the source of the poor aerodynamic performance. A rotary wing, on the other hand, can offer nearly constant lift throughout its turn. The inherent technological complexity of flapping wings is another limitation. Flapping wings must provide lift force while also contributing to pitch, roll, and yaw control in flight, making an automatic pilot incredibly hard to create. Furthermore, the notion that rotary-winged drones did not originate as a result of the evolutionary processes of living processes should not dissuade experts from considering rotary-winged drones as feasible VTOL alternatives. Undeniably, despite their high efficiency, neither wheels, propellers, nor rotors have indeed been manufactured by the usual course of progression. Nature, according to the researchers, has a few specific exemptions to this lack of creativity, such as whirligigs (helicopter) and the hairlike attachment in bacteria, which is called a flagellum. The whirligigs on the maple, by contrast, are essentially an active rotary-wing flyer or helicopter kind that, when released by its own parent, uses its higher glide proportion to reach far regions. Influenced by the whirligigs of helicopter flying, a mono-armed nano drone was developed, which resembled everything the same as the influencer, such as its weight, flying style, etc. Due to their ability to mimic bugs and flyers of nature, which can effortlessly vanish from human eyes, fluttering-armed drones became very useful in defining and recognising missions requiring stealth in the long run. Novel modern resources, like size and shape-changing polymers related to various types of MEMS, may also benefit fluttering-wing UAVs [9]. Furthermore, it is now much easier to create an insect or even smaller types of drones using newer fabrication techniques such as miniaturing, doping, deposition, and so on [10].

2.1.3 Mini Helicopter Drones

Apart from the limitations of permanent armed drones and the difficulty of fluttering-armed drones, helicopter-like drones piqued the scientific community's interest. According to recent figures, approximately 80 per cent of the approximately four thousand drones flying in the European country France are registered, with the majority of them being mini helicopter drones. The main reason for this interest is that helicopter drones have the ability to hover. That will make them more distinctive than the others. Because of its special characteristics for motionless flight, this drone will have no trouble taking off and landing. Since it is able to hold its position over a defined reference area, pre-preparation of the landing path won't be required. Having seen the other side, these kinds of drones can be easily designed and constructed and can be used for more domestic applications. A decade ago, only drones with four rotors were available, which are multiple rotors, but nowadays drones with different combinations of rotors are possible, which are hybrid ones. These hybrid motor drones exhibit more robustness and reliability than other drones. The motors in this drone are positioned evenly in the azimuthal coordinates. But it is not a mandate to follow the defined coordinates. Many designers use different coordinates and elevation methods, which might give a better line of sight beyond the drone's front eye range. One such example (Figure 2.8) available on the market now is the Falcon Version produced by Ascending Technologies.

The Version exhibits the upgradation of features in the copter.

Whilst the main rotor and an anti-torque rotor are combined in a helicopter, flight control is additionally provided by a cyclic-pitch swashplate. Therefore, compared to creating multi-rotors, designing a helicopter therefore takes far more experience and expertise. Reynolds effects cause the propulsion efficiency to decrease as the rotor diameter is decreased.

FIGURE 2.8 A mini helicopter with eight rotors created by Ascending Technologies. (Image courtesy of Lakeside Labs GmbH.)

FIGURE 2.9 1 kilogramme mini-heli drone. (Photo courtesy of Ascent.)

It is much more effective to employ a single rotor to its extreme instead of using several rotors, each with a smaller diameter, to cover the same disc area. (Figure 2.9)

However, it is also possible to use a tail rotor as in traditional planes, and one can also include an extra coax motor below to counteract the resulting torque. Such a coaxial rotor enables vertical axis control and altitude hold. Figure 2.10 is a standing example of a drone with gimbal support, which will allow it to rotate in any direction. This kind of drone can hover for up to approximately twelve minutes and will become a backpack once the blades of the drones are folded.

2.1.4 Translatable Drones

The ability of the multiple motor drones to perform adequately in windy outdoor conditions has hampered their success. A multitude of adverse airflow effects, such as poor motor efficiency when the entering freestream is strongly slanted with regard to the rotation axis, impede elevated onward flying. Whereas permanent wing arms are unable to produce proper drift, mini helicopter drones are mostly slower in nature

FIGURE 2.10 ISAE-SUPAERO created a vented coaxial rotor microdrones having less than a kilogramme. (© Aéroland. Reproduced with permission.)

and are typically much less efficient in high-speed flight phases. As a result, most of the drone's designs aim to combine the benefits of permanent arm and mini helicopter drone designs. Convertible UAVs are these particular design fusions. The advantages of rotary- and fixed-wing layouts can be combined using two different design approaches. One method involves starting with an aeroplane configuration and modifying it to enable vertical flying. The other approach involves starting with a mini helicopter drone formation and varying it to attain flat level flight. The Flex rotor is a member of the family of tail-sitter or tilt-body unmanned aerial vehicles (UAVs), which can travel horizontally and execute vertical take-off and landing. The wingtip rotors' folding blades provide for a reduction in drag when in aeroplane mode. When flying in helicopter mode, the huge propeller that powers the aircraft simultaneously functions as the main rotor. Mostly because of this, the pitch should be adjusted based upon the movement of the drones. As they are flying, the pitch should be adjusted to the maximum, and vice versa. The convertible biplane design, which combines a typical multi-rotor structure with a stack of uplifting panels placed underneath [11], serves as an illustration of the second tactic. The crucial distinction here is that the body tilts at a 90-degree angle when flying horizontally. When in aircraft mode, the vehicle operates with good aerodynamic efficiency as a permanent hovering arm drone. The absence of a horizontal tail may prevent the flying wing from being statically stable, but the motors attached to the wings can be employed to adjust pitch. Recently, the France-based drone producer suggested a marketable type of such a permanent wing aircraft with the idea of a tilt-body that uses a cross-wing other than the typical permanent wing aircraft (Figure 2.11). UAVs with tilt-rotor and tilt-wing layouts are more examples of convertible designs. The main body of tilt-rotor UAVs can stay horizontal when changing from cruise to hover by putting the rotors on a revolving axis. In the case of tilting drones, during flight both the rotor and a part of the arms that is actually in the slipstream range of the propellor will rotate together as both are attached to each other. Such convertible aircraft call for an auxiliary tilting gear in both scenarios, which adds weight and complexity. Additionally, the positioning of the overall centre of gravity would change during transition with movable components, such as motors, which makes creating the autopilot more challenging. The RWTH Aachen University's AVIGLE is an illustration of a tilt-wing UAV [5]. The wing of the AVIGLE UAV may be tilted upright while the fuselage remains horizontal, despite the fact that it has the appearance of a standard airplane. It should be noted that in order to preserve pitch control during transition, a second vertical rotor must be placed close to the tail. The Skate, an economical and adjustable UAV designed by Aurora Flight Sciences, is a good example. The Skate is a rectangular aircraft with separate roll and pitch control provided by two electric motors mounted on separate tilting systems. A differential throttle provides yaw control. Therefore, no additional moving components, such as elevators or flaps, are required to operate the vehicle.

Because the drone is angled steeply in flight, the rotors almost align with the wing chord in both cruise flight and hover.

The rotor axis only needs to tilt with respect to the wing during transition flight. However, with some tilt-rotor arrangements, like with the Osprey V-22, the fuselage must remain horizontal. The primary advantage of using this tilt-rotor layout is that its systems, which are both hardware and software oriented, antennae, and consignment position do not need to be changed to accommodate an attitude shift while hovering. However, because the propeller slipstream impacts a section of the wing, tilting the rotors typically results in a down force.

Finally, two other configurations that are only marginally relevant to UAV designs should be mentioned. Let's get ourselves straight into the design structure of the modified parachute called a paraglider. The body of this improved parachute is made up of a propellor on one side and the parachute itself on the other part of the body, which flies high based upon it. The Swan, a UAV of this type created by Flying Robots in Switzerland, serves as an illustration (Figure 2.12). The ability of a glider UAV to fly

FIGURE 2.11 Aerovel Corporation (Courtesy) created a 20-kg convertible UAV.

FIGURE 2.12 Left: Flying Robots, Switzerland created a paraglider UAV. (Left picture credit: Flying Robots, Swiss.)

Right: Feather like Drones. (Right photo courtesy: Ride Engineering, Russia.)

extremely slowly and be packed very compactly is its greatest benefit. For search and rescue efforts that span broad areas, since they can be launched and lowered from an airplane, paragliders are a fantastic choice.

Another or second configuration can be the airship. The Sphere-P2 project, created by Ride Engineering, serves as an illustration of one such approach. While the ability to hover for extended periods of time makes lighter-than-air UAVs appealing, they have two significant drawbacks: (1) their susceptibility to breeze, and (2) the inhibited freight which will be carried for a defined drone capacity. For continuous area observation, certain drones should be kept within a fixed distance. In the project named spearP2, a rotating or moving motor was created to offer drones height holding, and a changeable centre of gravity is used to enable horizontal control.

2.1.5 Hybrid UAV or Drones

A new kind of unmanned aerial vehicle (UAV) that has lately appeared for extremely practical reasons deserves special attention today. Drones with a recognition mission cannot escape running into unforeseen impediments of any kind when flying close to the ground, whether in a forest or an urban environment, like tall buildings, large transformers with crossing electrical lines, smokestacks, feelers, and many more. Additionally, some identification tasks might require vehicle entry through very small tunnels or corridors, as well as building intrusion. Obstacles in such mission profiles cannot be avoided. The difficulty and danger of hopping over barriers may still severely limit the use of conventional ground vehicles. Additionally, it may be preferable in many situations for the UAV to land in the middle of the operation. For example, a police operation can unexpectedly call for a drone that is completely noiseless, like switching off the mechanical device in drones. Then the device must then touch down or stick to the ground. Nonetheless, it should have the ability to leave the ground and start flying with no assistance from a human. Vehicles that combine the characteristics of airborne and ground devices are known as "amalgamated drones." With these drones, impediments are no longer seen as issues to be solved, but rather as chances to include some new functionality. Design-wise, including an outside collision-proof framework and a pair of pultruded rods or hyper-eutectoid, may result in a heaviness forfeit but may be enhanced with new features like the ability to roll or hang from a ceiling. The MAV "Roll & Fly" is a four-sided airborne drone driven by two propellers that rotate in the opposite direction and is completely in a covered area, along with a couple of wheels on both sides of the drone arms, as a first illustration of such a hybrid vehicle. The vehicle's two elevators, which are placed in the propeller slipstream at the trailing edge of the wings, allow it to fly vertically far away from obstructions. The MAV can also fly with conventional bi-motor flying wings horizontally. The elevators, which are effective throughout the whole flight domain, enable control of pitch and roll in both scenarios. A differential throttle provides yaw control. The wheels not only shield the propellers when they collide with a flat surface like a floor, ceiling, or wall, but they also enable them to roll away from the wall at a consistent distance. When rolling on the ground, differential throttle can help the car "drive." Two further intriguing designs that follow the same principle of fusing ground and aerial vehicles should be addressed here. Both of them are predicated on the notion that the flying vehicle may freely rotate around the outside crashproof structure. In Figure 2.13, left, it is the representation of a drone which will have the capability of both flying in space and ground competences. It can easily bypass any obstacles on its way while it is on the ground.

The drone uses far less power when moving on land than it does when moving in the air and can easily navigate obstacles by just flying over them. The GimBall, created by a Swiss firm, serves as the second illustration (Figure 2.13, right). The airborne vehicle of the GimBall is housed inside an orb which will be able to move around independently between two axes such as an X (horizontal) and Y (vertical). Because of this, the drone will be able to go through any challenging terrain, such as a forest or a web of cables, without getting trapped. As a deduction to the section above, it seems that state-of-the-art technology has fundamentally changed how drones are typically categorised, making it impossible to reduce them to rotating arm or permanent arm drones. It is necessary to include cutting-edge configurations, including convertible and hybrid UAVs, in a broad overview of UAV concepts. For the aim of drone networking, the usage of portable and fusion or mixed drones is thought to be of the highest standing because they pave the way for multi-tasking missions that call for cooperation and dynamic job allocation.

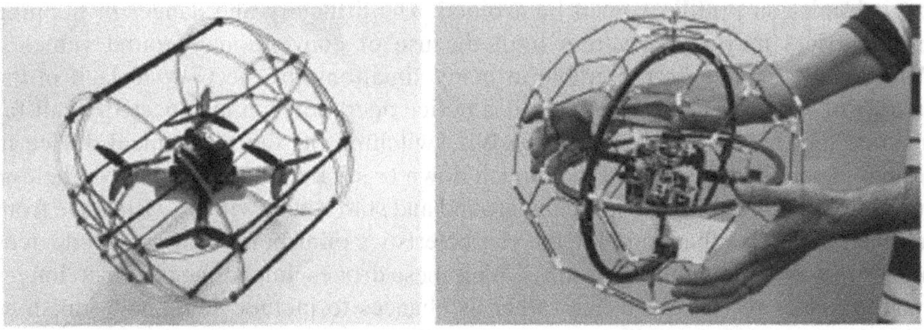

FIGURE 2.13 Left Image: Drone with both ground and aerial skills. (Image courtesy of Prof. Matthew, Chicago, Illinois.)

Right Pic: Dual-axis rotating sphere on the small unmanned aerial vehicle GimBall. (© Alain Herzog. With permission, reproduced.)

2.2 DRONE TEEMING AND DIMINISHMENT

Unmanned aerial systems (UAS) development has several excellent practical benefits, one of which is just economic. Heavier drones will be able to outperform traditional systems such as lighter drones if a specific shadowing or human-finding job can be completed economically. This also holds true inside UAS between bigger and smaller UAVs, where the modest scope of each and every drone will become a different cluster where all the groups will be working together. Even though practically any vehicle can be used to connect with drones, this is only possible for smaller drones. In fact, only smaller drones can be deployed quickly because they need a very small logistic footprint and a small crew. Therefore, launching dozens of UAVs in order to achieve a synchronised flight would be impossible if each UAV deployed took more than a minute. If not, the mission of the first airborne UAV would have ended and the mission of the last UAV would not have begun. Only tiny UAVs might be suitable for networking several different vehicles. A multi-vehicle surveillance mission can only be carried out using mini-or micro-UAVs. In order to operate a bigger drone available on the market, it entails human intervention a lot. As a result, the number of drones and their size will be indirectly proportional to the number of humans deployed in the work of drones as operators. This is the concept of drone teeming. Drone networks are created in a way that they will operate in clusters, which will reduce the number of humans involved. Ultimately, there is no need to operate drones individually, which takes lots of human resources. In order to efficiently operate the drone network, an armada of drones needs to be managed by a sole operator who should have a higher degree of independence, which might require and lead to innovative control and navigation algorithms. In the chapters that follow, these new algorithms will be covered in more detail. For the time being, it is crucial to concentrate on the practical challenge of managing a herd of drones that requires the ability to miniaturise every drone in a case in which faults or crashes of one drone do not create significant heavy damage in any practical cases and can be a highly fault-tolerant system even after it allows jobs assigned to be successfully finished. Therefore, before moving on, it is crucial to carefully assess how much drones can be trimmed down in the event of their size.

2.3 TECHNICAL CHALLENGES AND OPPORTUNITIES IN DRONES DIMINISHMENT

If the capacity to miniaturise aerial vehicles is continuously relied upon by UAV networks, miniaturisation itself suggests various potential as well as new design obstacles. Opportunities for reducing the visual and electromagnetic impact of UAVs include miniaturisation. Drones, which are smaller in e-aerial vehicles is continuously relied upon by UAV networks, miniaturisation itself suggests various potential as well as new design obstacles. Opportunities for reducing the visual and electromagnetic impact of UAVs include miniaturisation. Drones, which are smaller in size, will subsequently provide a greater benefit in terms of surreptitiousness for several defence and security-related applications. Smaller cars also typically make less noise and are less conspicuous when appropriately camouflaged to blend in with their surroundings. The ability to fit into extremely small areas, such as subways, derelicts, airducts, small passages, and drainage tubes, is another benefit of downsizing drones. Non-flying or large vehicles tend to be grounded in such constrained areas other than the vehicles that have flying capability. Finally, vehicles that are smaller in size are affordable. While hundreds of flying robots are pursuing the identification goal, losing 99 dollars to flying robots is not a major problem. However, losing a predator-sized unmanned aerial vehicle might have serious consequences for the operator. Last but not least, in some cases, combining or integrating available drones, which hope to be cooperative, might provide the best way to complete most byzantine tasks and will get more hands of drones, whereas if there is only one drone available, it will take more time and effort to complete a single task itself. Drones have several design and technological obstacles despite being highly desirable, including, but not limited to, dynamism to their surroundings, energy supply, and aerodynamic efficiency.

2.3.1 Sensitiveness to Dynamic Changes

It is not sufficient to simply scale down standard aircraft layouts when designing mini-UAVs. This is due to a number of factors. One has to do with how sensitively the vehicle perceives gusts. Let's take a look at a traditional fixed-wing aircraft to demonstrate the impact.

2.3.2 Density of Energy

Even though gasoline has a higher energy density than the greatest batteries now on the market, thermal incineration engines lose their efficiency when their size is dramatically decreased. The cause of this occurrence is that while the heat flux dissipated through the combustion chamber's walls only decreases as L2, the heat produced there is proportional to L3. Miniaturising thermal engines will thus invariably result in low thermodynamic efficiency because the majority of the heat generated inside the chamber where the ignition takes place will quickly escape, which can't be contained by physical barriers like walls or any objects etc. Due to the restrictions on chamber dwell time, increasing the speed of rotation takes place, which will offset radiation and also it will not provide a workable result. When thermal incineration engines are made smaller, additional issues like underprivileged compression tautness and increased abrasion also damage their appeal. Therefore, electrically driven vehicles—whose maximum value for a first-class battery is roughly 720 m^2/s^2—are the only option available to mini-drone builders. Energy density continues to be a noteworthy blockade to further miniaturising drones at this time, despite incredible progressions in the realm of batteries and other innovative chemical cell substitutes.

2.3.3 Flying Efficiency or Performance

The driving force behind the reason of glutinous properties in the drift turf around a flying drone helps in identifying a pattern based on different fluid situations, which may be indicated as

$$R = \rho \upsilon . len / \mu \tag{2.1}$$

Where,
ρ = compactness of the liquified solution
υ = swiftness of the solution
μ = gluiness of the solution used
len = length of the solution used.

In other words, it is the ratio of inertial force to that of viscous force. Low Rey numbers increase the likelihood of laminar separation, which lowers extreme highs and increases slog, though the drones are even flying in a very low range. The flying efficiency of mini drones will be limited by having critical design issues that demand new flying techniques for take-off efficiency with the least amount of drag. This is because the flying performance of drone arms surfaces and also that there is a possibility of degradation of drone blades based upon some driving force.

2.3.4 Undiscussed Design Obstacles

The nano-sized drone's difficulties not only exist in its design properties but also seem to exist in terms of practicality too. The problem with drone making is its size itself. As the working gears are joined together to be worked in a smaller space, every component inside the drone will be easily radiated as much as that. Besides, the working gears inside the drone need to communicate as well, so in order to do so, the electrical wires are used up. Those wires will also be accountable for the extra weight of the drone. Drones must be manufactured with lighter weights. Hence, amalgamation is required to condense the weight produced by the communication wire among the working gears of the drone.

2.4 DRONE'S NETWORK AND THEIR BENEFITS

In a flock or network of drones, each drone functions as a vertex, relaying information intended for other nodes in the network or communicating its own information to other nodes. This network can be thought of as a flying or mobile wireless network which may be supported by a base station, may be a satellite-based communication set-up, or it may be with no auxiliary substructure.

The design or topology of the above-mentioned drone's network could vary depending on the project and use cases or its systems scenario. Let's first examine the necessity for such a drone network. One benefit of a solo drone is that when it is at a higher loftiness, it will be a direct beeline into the system of a network, where every component involved This is the motivational reason for escalating several antennas or cell towers to nearly equal heights at a particular height of around 200 feet. Any single drone in the network can be a vertex itself between the components involved in the system of networks where any transmission can be sent through the vertex itself if it is in the network itself. This kind of drone wireless network set-up will provide a high-end infrastructure for effective communication to take place as the earth's environment is not even a completely flat surface, which will be actually a zig-zag surface itself. There are drones available that show how any duo drones will communicate with each other from the base station and vice versa. If only a greater number of vertices were included, the communication distance could be increased as much as possible.

The aerial mobile ad hoc networks can distribute information over longer distances in real time. Here, every individual node will be acting as an intermediator and also an end point, which sends the data for the entire network. All these individual nodes or drones can be collectively called a swarm of drones, which also avoids the need for supplementary set-up for communication. But the base station and space capsule-based services could be an added advantage for reliable and robust communication. The need for geographical data will be attained through the global positioning system sensor, which will be transmitted to every vertex in the network infrastructure. An airborne network is a general term for a drone network that uses both earthly-and beyond-atmosphere-based communication equipment.

2.4.1 Characteristics of Airborne Networks

As a result of aerial nodes moving quicker than ground-based nodes, an air system's architecture can be extremely dynamic. Specific methods for routing and safe information sharing are needed due to the dynamics' high rate of change. In order to ensure that the terminals hold a minimal innocuous distance throughout the flying time, we might be in need of recognise-and-eliminate and context-aware algorithms may be required. In many ways, vehicular networks involving solely ground vehicles and airborne networks are dissimilar and unique. Traditional security and mobility paradigms created for mobile adhoc networks and normal vehicle networks are not appropriate for fly-adhoc networks (FANET). For airborne networks, agile frameworks that account for the particular features, such as advanced information assurance, authentication, and integrity verification techniques, are necessary. A cyber-physical system (CPS) called an airborne network has active interaction between its physical and cyber components. Flight routes, manoeuvre geometries, and multi-mode resources, including ground-based nodes and control stations, make up the CPS's physical components, while compute, communication, and networking components make up the system's cyber components. The primary challenge for airborne networks is to achieve synergistic interactivity between their cyber and physical components. The next generation of air transportation systems will greatly benefit from this synergy if it is successfully explored and utilised. For instance, predicting the trajectory of airborne vehicles within the neighborhood (let's say a 1,000 square mile region), creating a trusted network with ally nodes, reconfiguring the network as its topology changes, and securely sharing audio and video streaming data among the pilots will greatly enhance situational awareness. However, basic design concepts are still lacking to explore this symbiotic relationship between the cyber and physical dimensions. The creation of experimental datasets is essential for the development of the above-mentioned design ideas.

2.4.2 Drones Network Agile Framework

The Agile framework offers a method for conducting connection research, assessing the performance of networks, and ultimately creating dependable routing algorithms. The Agile framework in particular catches the arbitrarily pattern of manoeuvre of every network model intermediator, where we acquire extensive data about every system architecture used, like dissemination of nodes in the network and connection track and lifespan statistics, which may be inferred. It is essential to create plausible and manageable mobility models for aerial networks, which do precise forecasts, which can be helpful in aerial networking. Several agile frameworks, such as random direction (RD) and random waypoint (RWP), have received substantial research in the last decade. The RWP model presupposes that an agent chooses an arbitrary point or place (a waypoint) and its driving pattern, and that its hiatus after reaching the end point before it can go to another agent. The extension of the defined RD model infers that an agent, after an arbitrarily chosen travel time, also selects its driving style and route at random. The literature contains information about these available frameworks' stochastic characteristics, such as their altitudinal dissemination. The popular RWP and RD models are ideally adapted in order to capture the ad hoc mobile user behavior in MANETs. They are unable to characterise characteristics specific to aerial vehicles, nevertheless. As an illustration, mobile people and ground-based vehicles can quickly slow down, turn sharply, and move in the other direction (look for an upgraded arbitrary agile framework that can capture these actions). Aerial nodes are unable to quickly reverse their direction of movement or make such acute turns. Therefore, it is necessary to create accurate models that include characteristics specific to airborne networks.

2.5 TECHNICAL ISSUES

When employing UAVs in the drone-base terminal environment, we must take into account network performance characterisation, the optimum drone deployment strategy in three dimensions, the allocation of wireless and computing resources, flight duration optimisation, and network design. Difficulties including channel modelling, control transfer management, time delay control, interference management, and 3D localisation should be taken into account during operations in an unmanned aerial systems scenario. UAV capabilities, flight restrictions, energy limits, and flying altitudes must all be taken into consideration when using UAVs for certain modern wireless communications services and their applications. For maximum performance and effective engineering, precise channel characterisation is crucial, while UAV communication has unique performance parameters of its own. Regarding the causes of altitudinal and time-based fluctuations around the non-stochastic drone channels, the propagation properties of swiftly changing drone streams have not yet been investigated. The construction of a general modulation scheme for drone air-to-ground communication necessitates extensive mock-ups and calculations in a variety of settings. should be considered for the drone's flying heights, antenna motions, drone shadowiness, etc. Although early investigations were conducted by Sun et al. in early 2017 on airframe shadowing, which is otherwise an obstruction or blockage induced by tiny rotary UAV building systems and maneuvering, further research into this issue is still needed. Figure 2.14 depicts the design issues encountered by the UAV-assisted communication design.

2.6 RESEARCH DIRECTIONS

If truth be told, the inimitable advantages of UAVs have found many applications, particularly in the wireless field, but safety is always a higher concern on the research side. Due to the applications in many fields, the system design of UAVs takes its newest form every time, which makes security-oriented UAV research its most copious form.

UAV communication channel design: Because UAVs discovered a plethora of new applications to implement, there will be no standard or common communication model that can be applied to any situation or application. The physical layer security influences channel design characteristics, which also have an impact on the information or data that is in reception of the network. Recent research shows that there are diverse methods available to do processing of signals, transmission controlling, coding or programming techniques, and most liable GPS jamming systems. That holistically makes the physical layer security of UAV the most efficient one. Also, PLS will be able to provide excellent security mechanisms against unauthorised access. Other than that, in technical aspects, the PLS is low cost for packets transmitted and its computations, which are involved in the network, thus making it applicable for IoT devices and UAVs.

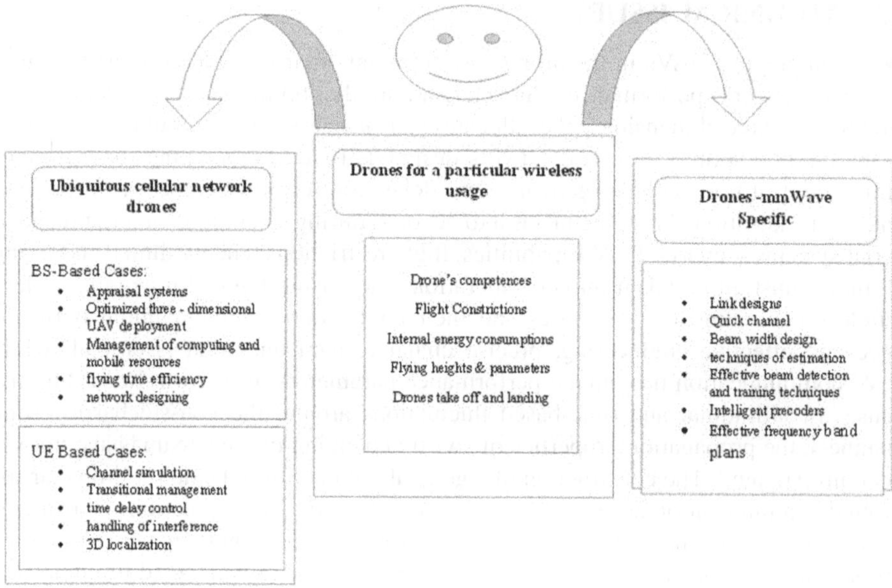

FIGURE 2.14 UAV-assisted communication system design challenges.

Beamforming design: Though various new beamforming designs keep coming into the market, such as position based, time division multi-array, millimeter wave, dynamic active-passive, 3D, location aware, learning based predictive, adaptive, the framework design seems to be challenging still, and the elevation adjustment tends to be somewhat critical based on the application.

Joint optimisation design: The resource allocation, the communication system, and the safety issue of the UAV flight mechanism have been resolved and handled by the optimisation design, in which it only produces a local optimum solution, whereas the profound inference will be the missing part in the result analysis.

Available energy in the system: Since the energy storing capacity of UAVs is limited in nature, it is always hard to concurrently progress in the case of energy performance and reliability of the system. The performance of the system related to energy will be mostly allied with unmanned aerial vehicles, which are completely dependent on their gesticulation features, such as their variety or range of speediness. To be exact, the design of the UAV trajectory is much more byzantine. Additionally, UAVs can be extended by utilising New Transmission Technology (Wireless). On the other hand, more concentration has to be given to the security aspects and performance of the UAV, as the signal itself becomes an interference, which could be detrimental to the quality and accuracy of the UAV's mission.

Trajectory optimisation: By maximising the UAVs' trajectory, wireless networks with UAV assistance may achieve much higher throughput and coverage levels. This

TABLE 2.1
Cutting-edge Techniques for Improving Drone Trajectory

Research paper reference	Optimised parameters	Enhanced parameter effects on the drone and its system
[12]	Drone's user planning and gliding path	Increased regular data transmission rate
[13]	Drone's several antennae along travelling path	Uplink network communication rate was increased
[14]	Together maximised trajectory and link transmission range	Increased throughput
[5,15]	Trajectory tracking algorithm	Less energy consumption
[16]	Drone's trajectory with optimised linear functions	Fuel usage reduction
[17]	Trajectory tracking	Probability of finding the objective
[18]	Drone flight path	Wireless adhoc network connectivity has improved

optimisation is influenced by issues such as flying limitations, energy consumption, drone user requirements, crash detection and evasion, accurate channel design, drone agility, and so on. Table 2.1 summarises the effort made so far to improve the efficiency of drone systems by creating the optimal drone flying positions.

Author Chen suggests a fully independent drone that itself adapts to the system entirely based on continuous monitoring.

2.7 AMALGAMATION TRIALS OF DRONES OVER 5G

Airborne telecommunication and connectivity of drones provide a number of issues that must be properly investigated. For example, dependable and low-delay communication is critical for effective drone operation. Present mobile networks are largely intended to provide increased broadband access to domestic consumers. Furthermore, regional landscapes with restricted grounded access point (GAP) service may not offer the essential connection services to connected drones, necessitating realistic solutions for the effective utilisation of this technology.

2.8 FUTURE STANCES

In previous sections, we discussed compatibility issues and identified potential 5G improvements to overcome a few of those constraints for connected drones. Notwithstanding countless research, there remains a vast number of outstanding challenges which must be studied. The following subsection attempts to highlight potential prospects for scholars as well as throw light on intriguing vacant subjects.

2.8.1 Unloading Computations

Owing to drones' low onboard power and poor processing capability, edge computing might be useful for unloading computation-demanding activities from connected quadcopters onto edge units, improving drone endurance and long-term career. Facial recognition in a mass observation application case is an illustration of a demanding assignment. By combining edge computing and drone competence, the detection work may be unloaded in order to finish the project on schedule. Furthermore, adequate security procedures must be implemented into the edge-based drone platform for enhanced efficiency in order to protect the data from eavesdroppers. Such scientific disciplines have mostly gone untapped thus far, and there are tremendous opportunities for designing and developing edge-based frameworks for connected drones given their wide application.

2.8.2 AI and Machine Learning Approaches

AI and machine learning-based techniques were viewed as powerful tools for addressing a wide range of wireless network issues and transforming 5G infrastructure. With the introduction of connected drones, a new level of innovation for independent aircraft operations in regard to security, efficiency, and ever-changing multidimensional situations has emerged.

There are several experimental difficulties in exploring and analysing AI/ML-enabled solutions to handle obstacles such as unmanned aerial congestion control, transmission energy administration, inter-drone collaboration, and the like. Furthermore, model-free reinforcement learning aids drones in addressing safety problems by dynamically controlling transmission based on the harmful kind of threat.

2.8.3 Energy Constraints

Regarding drones, internal power is indeed a constraint. Technological breakthroughs in portable power packs and the utilisation of solar energy are two methods for extending the flight life of drones. Drones necessitate a steady battery pack to function since they consume a large portion of the power when airborne. Streaming, receiving, application feature operation, and route planning enhancements all require a power pack. Many recent studies on connected drones don't really take this power restriction into account. There is indeed a significant amount of space to examine the efficiency of connected drones given the limiting energy limitations, notably in the field of virtual network implementation on drones, enabling automation, flight path optimisation, real-time learning approaches, and task lifetime computation.

2.8.4 Network Design and Cell Planning

Drones are often serviced by the sidebands of existing cell sites in 4G or 5G wireless connections, allowing for easy cell choice with distant base stations. Additionally, the rapid agility of drones boosts radio connection losses owing to far more rapid handoffs. Telecom companies and device manufacturers need transmitters that can handle hovering drones at higher elevations as well as improved methods for radio cell planning.

2.8.5 Safety and Confidentiality

The interconnected drones were subjected to malware or illegal attempts to counterfeit command signals because of the free or exposed communication channels. These efforts represent a significant risk to the drones in terms of leakage of classified material or operational failure.

Duty cycle counterfeiting could have a negative impact on the drone operation and make it harder to restore its functions. To prevent such hostile alterations, a key ongoing topic is to improve the safety and confidentiality aspects of drone mobile communications, which necessitates deeper research into security considerations across all tiers of communication layers.

2.9 CONCLUSIONS

Another more crucial step in the creation of future connected cities in the 5G-IoT era will also be based upon UAV-assisted wireless communication networks. UAVs have been dominating the skies for the past four decades and are essential to wireless communication networks. Numerous difficulties with this technology have been found by scientists from around the world, and workable solutions have been suggested. Also, a strenuous effort has been made to emphasise a handful of difficulties that must be solved in order to create the best assisted networks for the UAV system, clearing the way for future studies. Numerous research obstacles have been found and addressed in recent years, and this technology is progressing quite quickly. The development is still in progress.

ACKNOWLEDGMENTS

I'd like to thank Professor Dr Prasanna Venkatesan, Pondicherry University, my research supervisor, for his patient instruction, passionate support, and constructive criticisms of this study effort.

I'd also want to thank Professor Matthew, Chicago, Illinois, and Philip Richardson for their acceptance in recreating images, which helped in keeping my development on track.

I'd also want to thank the technicians at the department's laboratory for assisting me in obtaining the resources I required to run the programme.

REFERENCES

[1] Akoramurthy Balasubramaniam, Reshma S, Senthamizh Selvi P, Arthi Janardanan, Enhancing the proficiency of the drone with an application of ocean farming, *Procedia Computer Science*, 133, 725–732, 2018, ISSN 1877-0509, https://doi.org/10.1016/j.procs.2018.07.126

[2] Erdelj, M., Uk, B., Konam, D., Natalizio, E., From the eye of the storm: An IoT ecosystem made of sensors, smartphones and UAVs, *Sensors* 18 (11), 3814, 2018.

[3] Sun, R. and Matolak, D.W., Initial results for airframe shadowing in L–and C-band air-ground channels. *Proc. Integrated Commun., Navigation, and Surveillance Conf,* 1–8, Apr. 2015.

[4] Yaliniz, R., El-Keyi, A., Yanikomeroglu, H., Efficient 3-D placement of an aerial base station in next generation cellular networks, in: *Proc. of IEEE International Conference on Communications (ICC)*, Kuala Lumpur, Malaysia, May 2016.

[5] Hentati, A.I., Fourati, L.C., Comprehensive survey of UAVs communication networks, *Comput. Stand. Interfaces*, 103451, 2020.

[6] Lyu, J., Zeng, Y., Zhang, R., Cyclical multiple access in UAV aided communications: A throughput-delay tradeoff. *IEEE Wirel. Commun. Lett.*, 5 (6), 600–603, 2016.

[7] Abdel-Malek, M.A., Ibrahim, A.S., Mokhtar, M., Optimum UAV positioning for better coverage-connectivity tradeoff. *IEEE 28th Annual International Symposium on Personal, Indoor, and Mobile Radio Communications (PIMRC),* Montreal, QC, 1–5, 2017.

[8] Galkin, B., Kibiłda, J., DaSilva, L.A., Backhaul for low-altitude UAVs in urban environments, in *Proc., International Conference on Communications (ICC*), 1–6, May 2018.

[9] Li, B., Fei, Z., Zhang, Y., UAV Communications for 5G and beyond: Recent advances and future trends. *IEEE Internet Things J.,* 1 (1), 99, December 2018.

[10] Ahmed, N., Kanhere, S.S., Jha, S., On the importance of link characterization for aerial wireless sensor networks. *IEEE Commun. Mag.,* 54 (5), 52–57, May 2016.

[11] Mozaffari, M., Saad, W., Bennis, M., Debbah, M., Wireless communication using unmanned aerial vehicles (UAVs): Optimal transport theory for hover time optimization. *IEEE Trans. Wireless Commun.*, 16 (12), 8052–8066, Dec. 2017.

[12] Wu, Q., Zeng, Y., Zhang, R., Joint trajectory and communication design for multi-uav enabled wireless networks. *IEEE Trans. Wirel. Commun.*, 17 (3), 2109–2121, 2018.

[13] Jiang, F. and Swindlehurst, A.L., Optimization of UAV heading for the groundto-air uplink. *IEEE J. Sel. Areas Commun.,* 30 (5), 993–1005, June 2012.

[14] Zeng, Y., Zhang, R., Lim, T.J., Throughput maximization for UAV enabled mobile relaying systems. *IEEE Trans. Commun.,* 64 (12), 4983–4996, Dec. 2016.

[15] Franco, C.D. and Buttazzo, G., Energy-aware coverage path planning of UAVs, in: *Proc. of IEEE International Conference on Autonomous Robot Systems and Competitions (ICARSC*), 111–117, Vila Real, Portugal, April 2015.

[16] Grøtli, E.I. and Johansen, T.A., Path planning for UAVs under communication constraints using splat! and milp. *J. Intell. Robot. Syst.*, 65 (1–4), 265–282, 2012.

[17] Tisdale, J., Kim, Z., Hedrick, J.K., Autonomous UAV path planning and estimation. *IEEE Robot. Autom. Mag.,* 16 (2), 35–42, 2009.

[18] Han, Z., Swindlehurst, A.L., Liu, K., Optimization of MANET connectivity via smart deployment/movement of unmanned air vehicles. *IEEE Trans. Veh. Technol.,* 58 (7), 3533–3546, Dec. 2009.

3 Control Algorithms for Multiple Autonomous Underwater Vehicles to Cooperatively Encircle Intelligent Intrusion Devices

Hongfei Li, Mingzhi Chen, and Haoxuan Song

3.1 INTRODUCTION

71% of the earth's surface is the ocean, which is rich in minerals, biology, chemistry, and water resources. It is a treasure for human beings. The ocean has become the main battlefield for future competition. Smart devices may invade territorial waters to fetch relevant marine resource information. The establishment of a fast and reliable underwater target hunting system can effectively combat various equipment that wants to invade the territorial waters and strengthen the underwater defense system. An Autonomous Underwater Vehicle (AUV) integrates high-tech technologies such as sensor technology, underwater acoustic communication, navigation, artificial intelligence, and machine learning. It can operate autonomously, is flexible and lightweight, and is the best choice to encircle intrusion devices [1]. Multiple autonomous underwater vehicles (multi-AUVs) may form an effective water resource protection system by cooperatively and autonomously hunting down intrusion devices.

Due to the non-cooperative nature of intrusion devices, multi-AUV encircling systems usually belong to non-cooperative adversarial multi-agent systems. The AUV wants to encircle the intruding device quickly, and the intruding device does its best to avoid being encircled, preferably by making the encircling time infinite. On the topic of improving the success rate of encircling, many algorithms are proposed, which contain model-based and behavior-based algorithms, probabilistic prediction algorithms for successful encircling [2], etc. In addition, many improved algorithms have been derived, such as angle priority algorithms, rule-based (fuzzy rules), and reinforcement learning-based algorithms [3].

DOI: 10.1201/9781003320074-3

Initially, related research focused on encircling intruding devices without escape intelligence, and proposed algorithms based on model and behavior. The model-based algorithm has a high encircling efficiency and success rate [4], but it needs to construct accurate robot kinematics and dynamics models, which is difficult to achieve in the application of AUV. Behavior-based algorithms reduce the complexity of the encircling task by eliminating the need to build an accurate mathematical model of the robot, however, the encircling efficiency will be reduced due to the inability to predict the behavior and location of intelligent targets.

With the deepening of research, the research on encircling targets with escape intelligence has been gradually carried out. It is more difficult to encircle targets with certain escape intelligence, but it is closer to the actual encircling situation. For the targets with escape intelligence, taking measures to prevent them from escaping is an important guarantee. Multiple robots cooperatively form an encircling formation, blocking the escape path of the escaping targets, which can largely prevent them from escaping. Therefore, many researchers focus on the algorithm that robots use formation control to achieve encircling. Wang et al. proposed a method to mark the "dead point" of the path, improved the artificial potential field for the encircling task, and pointed out that there is an optimal circle, which can maximize the encircling efficiency [2]. Literature [5] uses the BNN model to solve the multi-robot encircling problem. The proposed algorithm can enable robots to efficiently complete their real-time encircling.

The above research focuses on the two-dimensional ground encircling scenario, however, the underwater multi-AUV encircling environment is special. In the research of the AUV cooperative encircling, literature [6] used multiple AUVs to eliminate mines and proposed a full-coverage algorithm, but the mines are static and have no intelligence. Literature [7] solved the cooperative encircling problem of multiple AUVs with a BNN model by sharing neuron activity. They proposed an algorithm to assign encircling tasks based on distance and negotiated through communication between AUVs, but sometimes conflicts were encountered between AUVs. To overcome this conflict, a position prediction algorithm was proposed in [8]. The algorithm can successfully capture the escaping target with a smaller sailing distance by comparing it with the artificial potential field method. However, they assume that the AUVs are faster than the evaders (3 to 6 times faster than evaders), which effectively reduces the difficulty of encircling [7, 8]. Afterward, to improve the encircling efficiency, the literature [9] proposed an encircling alliance generation algorithm based on time competition. Literature [10] completed underwater target rounding based on the spinal cord nervous system and genetic algorithm. However, their alliance generation method is based on the distance between the AUV and the task point and has nothing to do with the AUV's ability, so the alliance may not be reasonable. In addition, the path planning algorithm for encircling relies on the shared propagation of neuron activity. Because of the slow communication speed and narrow bandwidth in the underwater environment, it is impossible to guarantee the real-time transmission of so many neuron activities.

Based on analyzing the research status of multi-AUV cooperative encircling, multi-AUV cooperative encircling usually includes task allocation to form an alliance, and pursuit path planning to complete an encircling. Although there are many works focused on the encircling control of multi-robots or multi-AUVs, they usually focus on specific stages of the encircling process and do not comprehensively address the problem. This chapter comprehensively solves the multi-AUV cooperative encircling problem and focuses on the task allocation and path planning of the two stages in encircling control.

Using the time-based improved Self-Organizing Map (SOM) algorithm to assign tasks to the multi-AUV system and the path planning algorithm based on the optimized GBNN model, the current underwater situation can be fully taken into account instead of simply analyzing the shortest distance. Because in the shortest distance, the influence of ocean currents on the pursuit path of AUV in the rounding problem cannot be fully considered. In the following simulation content of the AUV encircling, the time-based and distance-based AUV running total distance comparisons were compared. It is not difficult to see that the results show that the alliance generation method based on the time competition mechanism makes the AUV capabilities of the encircling teams the same, and improves efficiency. More importantly, the GBNN algorithm has low computational complexity and is faster. It does not need to share neuron activity between AUVs and is more suitable for cooperative encircling in an underwater weak communication environment.

The main structure of this paper is as follows. Section 1 mainly introduces the research status of the multi-AUV system. Section 2 introduces the task allocation problem in the encircling problem of the multi-AUV system. This paper establishes a mathematical model for it and focuses on the improved SOM algorithm. Section 3 mainly analyzes the problem of path planning, and the method based on the GBNN model is mainly used for the pursuit path in the process of encircling. In section 4, through a comparison of simulation on two mechanisms: time-based and distance-based, a conclusion is drawn finally.

3.2 TASK ASSIGNMENT AND ALLIANCE FORMATION APPROACHES IN MULTI-AUV COLLABORATIVE ENCIRCLING

Because of the increasingly complex underwater operations, the ability of a single AUV to meet the requirements is poor. Therefore, the application of multiple autonomous underwater vehicle (multi-AUV) systems has become important, which provides an effective way to complete complex underwater operations [11]. A multi-AUV system is a group of AUVs designed to perform some collective behavior through which some goals that cannot be achieved with a single AUV become feasible [12]. The system has several foreseeable benefits, such as the ability to solve the complex task, reliability, and simple structure design [13]. Researchers have completed many studies on multi-AUV systems, including cooperative search, formation control, dynamic task assignment, cooperative hunting, etc. [14–16]. Among them, cooperative hunting is more comprehensive.

One of the most challenging problems in a multi-AUV system is how to optimize the allocation of tasks so as to improve the performance of the entire system [17]. Task assignment is critical for the co-design of the multi-AUV system [18]. It is the basis of multi-AUV systems and reflects the organizational form and operating mechanism of multi-AUV systems' decision-making level. The efficiency of task allocation directly affects the completion of tasks and is related to whether an AUV can give full play to its ability. Therefore, task allocation is critical and challenging.

3.2.1 Problem Definition and Model Establishment

Multi-AUV task allocation is to prepare and assign tasks to AUVs. It is supposed that a set of tasks $T = \{T_1, T_2, T_3, \cdots T_m\}$ should be assigned to a set of available AUVs $R = \{R_1, R_2, R_3, \cdots R_n\}$. The problem is to find an optimal allocation to make the assignment a set of tasks to a subset of AUVs [19, 20]. Each AUV $r \in R$ needs to execute a bundle of tasks $G \subseteq T$, and the total cost of a set of tasks can be approximated as a sum of the cost of a single task [21].

$$b_r(G) = \sum_{k=1}^{f} b_r(t_k)\{t_k \in G\} \tag{3.1}$$

f represents the number of tasks of the bundle G, $b_r(t_k)$ is the cost of task t_k, and $b_r(G)$ is the total cost of task bundle. Task allocation determines the division of the tasks into bundles and which bundle $G \subseteq T$ and each AUV $r \in R$ needs to receive.

Task allocation of AUVs is subject to some basic constraints, such as environmental constraints (ocean current, unknown underwater environment, obstacles, etc.), AUV constraints (communication loss, low battery, etc.), and task constraints (multi-agent task, hierarchical tasks, etc.). The purpose of task assignment is to finally achieve the completion of tasks with the least costs. Multi-AUV task allocation is to minimize the total cost under constraints [22], which belongs to the knapsack problem and is Non-deterministic Polynomial hard [23].

The tasks of the multi-AUV system are diverse, including marine sampling, sea area inspection, target search, interception, and hunting. Among them, most tasks should be finished in a limited time, so some articles put the consideration of time cost in the first place. Equation (3.2) can be the objective function based on the shortest time, where the last completion time that an AUV can finish a task is defined as c_i.

$$\min \sum_{i=1}^{n} \sum_{i=1}^{n} c_{i,k} \tag{3.2}$$

The traveling distance of an AUV to finish a task is also important, and some other articles put more emphasis on the total distance. Equation (3.3) is the objective function based on the shortest distance, where $D_{i,k}$ represents the i-th AUV's distance to the k-th task.

$$\min \sum_{i \in R} \sum_{k \in f} D_{i,k} \tag{3.3}$$

There are many constraints to the solution of the objective function such as the limited power source of AUVs and the ocean current, etc.

3.2.2 Organizational Paradigms

The System organization for task allocation problems is mainly divided into three categories: centralized system, distributed system, and hybrid system.

3.2.2.1 Centralized System

The centralized system usually needs a powerful central processing unit. The central processing unit monitors basic parameters, such as the relative position, battery capacity, and task status, and then distributes the tasks to the AUVs. Advantages of this type include the reduction of duplication of effort, and resources, savings of cost and time, and better coordination [24]. In theory, it can achieve the optimal solution of task allocation. However, large computational complexity in the dynamic environment can be a huge challenge for the central unit. The environment changes continuously in the dynamic environment, the central unit may not be so fast to obtain good task allocations which leads to very ordinary and even poor performance [25].

Figure 3.1 (a) is a schematic diagram of a centralized system. As seen in Figure 3.1 (a), this system relies on the central processing unit to successfully distribute the tasks, when the central unit fails, the whole system also fails. In addition, it needs fast and reliable communication in the centralized system, which is almost impossible in the underwater environment, so the reliability of the system is low. Therefore, the centralized system is more suitable for a small group of AUVs with a rigid connection network.

3.2.2.2 Distributed System

Unlike the centralized system, the distributed system does not require a globally connected network or a central processing unit. Figure 3.1 (b) is a schematic diagram of a distributed system. As can be seen from Figure 3.1 (b), several AUVs form a local team. AUV in the local group observes the state of neighboring AUVs and communicates with these AUVs [26]. The amount of information exchange between local teams is small because each AUV makes decisions by itself without global instruction [27].

Contrary to a centralized system, the main advantage of the distributed system is flexibility, robustness, and low communication demands [17]. It is so robust that if one of the units fails, the rest of the units in good working order can still work together normally. However, the consistency of task assignment is difficult to achieve in distributed systems [28]. It may not be able to obtain the global optimal solution, since it tends to have more local optimal solutions.

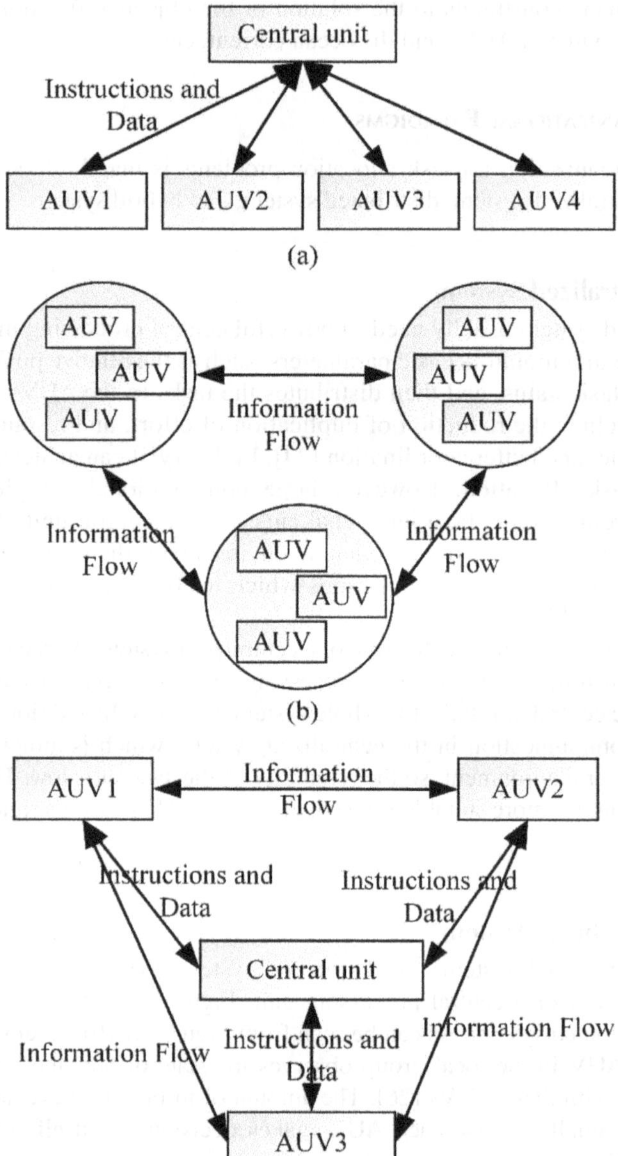

FIGURE 3.1 (a) Centralized system, (b) Distributed system, (c) Hybrid system.

Generally speaking, underwater systems cannot afford the high cost of establishing a centralized control center, and the current technical conditions cannot meet the needs of centralized methods to interact with a large amount of underwater information. Therefore, distributed task assignment methods are more suitable for underwater systems [29].

3.2.2.3 Hybrid System

The hybrid system combines centralized and distributed architectures. As shown in Figure 3.1 (c), the hybrid system reflects both overall decentralization and partial centralization. AUVs in this structure can also make task allocations by themselves and devote themselves to assigned tasks. Each AUV comprehensively considers its status, including energy, performance, etc., as well as factors task' as the task importance and difficulty, and finally, makes a decision to determine which tasks to perform. Meanwhile, a central unit continuously monitors the current status. In the event of a malfunction or inefficient allocation, the central unit can also issue instructions to avoid system failure and increase efficiency. To a certain extent, the monitoring of individual AUVs by the central unit is realized in the hybrid system, which is conducive to optimizing the solution. By adopting a hybrid layered architecture combining centralized and distributed systems, the shortcomings of the centralized system, such as low reliability and large communication load, can be overcome, and the flexibility and scalability of the system can be improved [19].

3.2.3 Dynamic Task Allocation Approaches

3.2.3.1 Market-based

Market-based task allocation is an outstanding multi-AUV task allocation strategy, which imitates the concept of market trading [26]. Figure 3.2 shows the basic process of auction-based task allocation. When the allocation of a task is triggered, all AUVs bid for the task, and the AUV that wins the bid will execute the task. The bidding

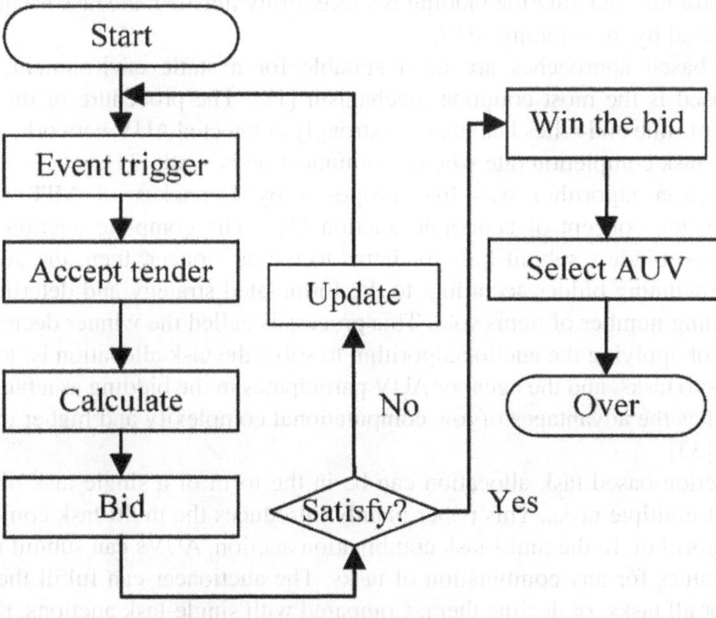

FIGURE 3.2 Auction-based task allocation process.

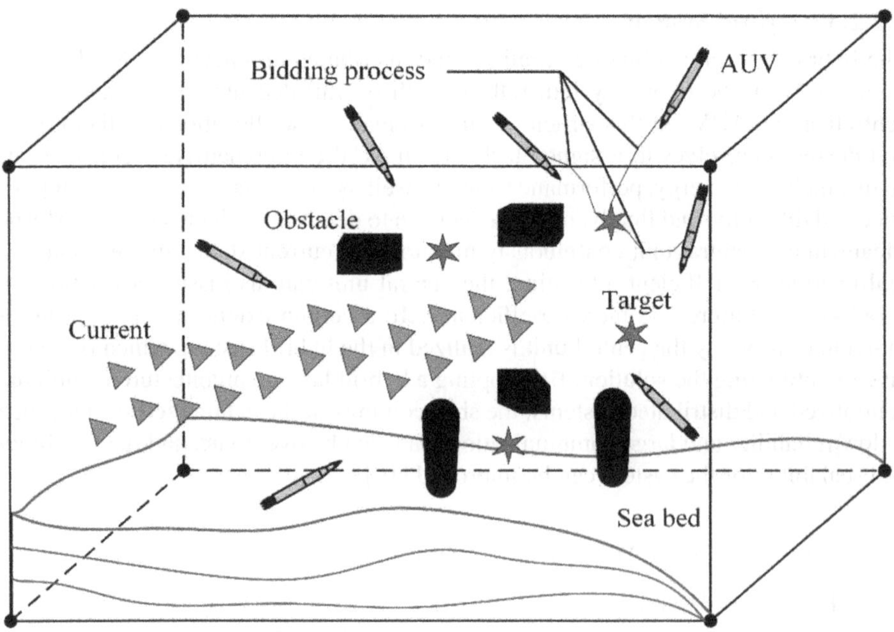

FIGURE 3.3 Bidding scenario for AUV task assignment.

scenario for the AUV task assignment is shown in Figure 3.3. A group of AUVs close to the target, bid for the tasks according to the distance [30], the income, energy, and speed constraints, and after the bidding is successfully finished, the task is carried out and completed by the winning AUV.

Market-based approaches are most suitable for a static environment, and an auction-based is the most common mechanism [18]. The procedure of the auction takes a lot of time and relies heavily on a strongly connected AUV network, so it will have a low task completion rate when communication is weak [31].

The auction algorithm was first proposed by Bertsekas of MIT in 1979, drawing on the concept of economic auction [32]. The complete auction process is as follows: bidders submit bids for items to the auctioneer, then, the auctioneer selects the winning bidder according to the formulated strategy and determines the corresponding number of items sold. This process is called the winner decision. The basic idea of applying the auction algorithm to solve the task allocation is: to correspond items to tasks, and the agent or AUV participates in the bidding as a bidder. The algorithm has the advantages of low computational complexity and higher operating efficiency [33].

The auction-based task allocation can be in the form of a single task or a combination of multiple tasks. This paper mainly introduces the multi-task combination auction algorithm. In the multi-task combination auction, AUVs can submit multiple total bid values for any combination of tasks. The auctioneer can fulfill the AUV's demand for all tasks, or decline them. Compared with single-task auctions, this form can better reflect the diverse needs of auctioneers and AUVs and improve auction efficiency.

The early auction algorithm was centralized. In the system, a centralized manager performed task auctions and assignments, such as the GOPHER designed by Caloud et al. [34]. Gerkey et al. proposed the MURDOCH based on a regularized, resource-centric, announcement, and bidding communication method. They used a sealed first-price auction to achieve task assignments [35]. Due to the high cost of establishing a centralized control center, at present, most of the distributed auction algorithms applied to the allocation of the multi-AUV system are distributed auction algorithms at present.

When applying the combinatorial auction algorithm to solve the multi-AUV task allocation problem, the two most critical problems are the AUV construction bid package problem and the bid winning bid determination problem. To deal with these problems [36], the author adopts an auction tree solution to the problem of winning bid determination in combinatorial auctions, aiming at the possible task deadlock phenomenon in multi-robot collection tasks. In [37], the author proposes the concept of the circular chromosome by improving the single-parent genetic algorithm, which solves the problem of winning bid determination in multi-task combination auction, and can achieve the real-time needs of multi-robot dynamic task allocation.

In [38], a resource-oriented decentralized auction algorithm is proposed, which assigns tasks to robots with the shortest completion time. In [39], an auction algorithm is applied to the task allocation in the multi-UAV system, especially considering multiple constraints.

However, in the above methods, the auctioneer does not fully utilize the market information, lacks certain flexibility in the strategy, and cannot maximize the interests of the auctioneer. In addition, the task assignment of underwater multi-AUV is compared with that of land. And, there are more serious uncertainties in multi-agent task assignments on the ground, including the influence of various environmental factors.

Aiming at ensuring the robustness of the task allocation system, Zhang proposes an innovative auction algorithm [40], which is a new method for determining the auctioneer and determining the bidding value of multiple AUVs, and also optimizes the auction algorithm to improve the security of task assignment and robustness. Aiming at the uncertainties caused by the underwater ocean current in the task assignment model, the author proposes a robust optimization algorithm to combat the uncertainties, which improves the efficiency of the multi-AUV system to handle complex underwater environments [41].

3.2.3.2 Intelligent-Optimization-based

Intelligent optimization algorithm, as a branch of applied mathematics, aims to focus on solving a specific problem to achieve the best solution among a feasible solution set. The objective function can be determined by optimization while meeting the requirement of the restrictions [42]. The common optimization algorithms include genetic algorithm (GA) [43], particle swarm algorithm (PSO) [44], ant colony algorithm (ACO) [45], and their variants. In the process of dynamic task allocation, there are usually numerous restrictions.

ACO algorithm [46], as a classic heuristic optimization algorithm, has been successfully used to solve problems such as traveling salesman. Through the improvement and expansion of ACO, the disadvantage that it is likely to fall into the local optimum has been overcome. Huge research results have been achieved in the research on the multi-agent task assignment method based on the ACO algorithm. Therefore, this algorithm is also adopted in task allocation of the multi-AUV system. The improved ACO algorithm proposed by the author improves the selection method and global pheromone update method of task execution ability ants, to improve the adaptive ability and global search efficiency of the algorithm [47]. The author adopts the dynamic ant colony's labor division (DACLD) mechanism [47], so that the overall intelligent behavior emerging from the interaction of many individuals with simple behavior can well adapt to the complex and changeable environment, and has a good ability to solve the flexibility of task allocation. However, through simulation of these, the influence of the environment on the completion of the task of the system is not taken into account, which lacks certain practicability.

Besides, the PSO also has a certain role in solving the task allocation of multiple-AUV. In 1995, Eberhart et al. proposed the PSO algorithm based on the research on the flight law of flocks and the behavior of biological flocks preying on flocks [48]. Due to the discreteness of the search space in the multi-AUV system, using the PSO algorithm presents a special problem since PSO initially operates in a continuous space. However, the efficiency of using the PSO algorithm is still relatively powerful, because it can solve the complexity of the multi-AUV task assignment problem well. Combined with the existing problems of PSO, scholars optimize PSO, and the proposed QPSO and CQPSO algorithm can solve the problem of task allocation very well [44, 49].

Apart from these, a rolling time-domain quantum artificial bee colony optimization algorithm was proposed to solve the multi-AUV task allocation, which realized the trade-off between the dynamic environment and cost of the AUV, and improved the task allocation performance of the multi-AUV system [50]. Literature [51] combines the improved simulated annealing algorithm and the transfer strategy to form a novel multi-objective local search alliance, which effectively solves the problem of multi-AUV task assignment.

The multi-objective optimization method will use less time to improve the task completion rate. Generally speaking, the task allocation rate of multi-objective optimization is higher than that of single-objective optimization. Usually, the basic objective function of a dynamic task allocation problem is completion time, AUV's moving distance, and completion rate. As far as the current research status is concerned, there is still an open research gap in the simulation and actual process of optimization-based dynamic allocation in the multi-AUV system.

3.2.3.3 Cluster Task Assignment

The basic idea of the clustering method is to take a certain characteristic of the task as a cluster of the cluster, and this attribute is an attribute that determines or has

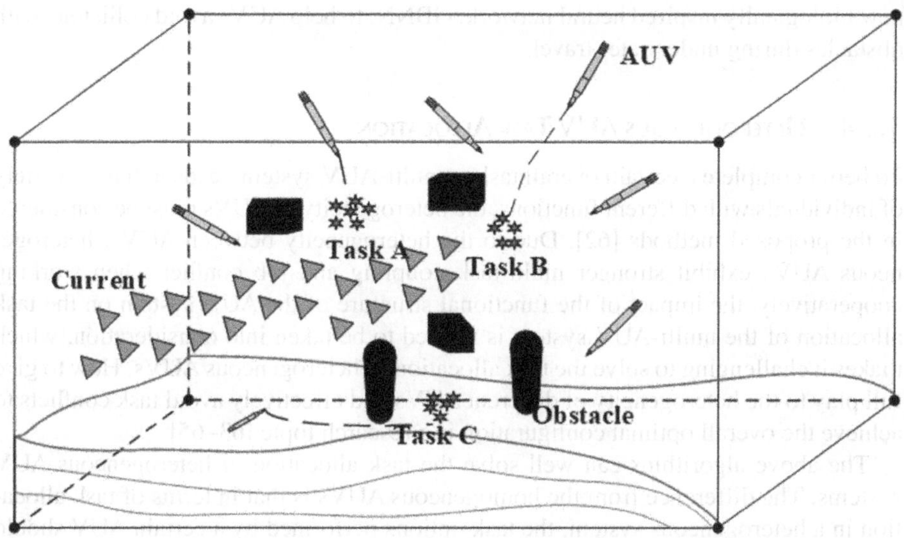

FIGURE 3.4 Divide similar or adjacent tasks into groups.

a significant impact on the allocation result. Although the result obtained by the clustering algorithm is sometimes not the optimal solution to the problem, the solution speed is fast, and it is suitable for task assignment problems with high real-time requirements. Cluster task allocation reduces the number of individual task allocations, so the computational complexity is greatly reduced. As can be seen in Figure 3.4, the system pre-positions some similar or identical tasks together, and then, AUVs can save the time of searching for the best tasks [52]. K-means algorithm [53–55], Self-Organizing Map(SOM) [51] [56–59], etc. are the most commonly used algorithms.

As a clustering algorithm, SOM has the following advantages: (1) the algorithm model is simple, and the amount of calculation involved is relatively low; (2) SOM does not directly find the optimal solution, which is very real-time. SOM was firstly used to solve the multi-task allocation problem for mobile agents in Zhu and Yang [60].

On the basis of this work, Zhu et al. [59] and Huang et al. [56] applied SOM for dynamic task allocation and path planning for the multi-AUV system. To balance energy and allocated workload, the modification was made to the winner selection function in the conventional SOM so that only the tasks assessed to have the highest and most realistic chances of achievement will be assigned to a single AUV. In addition, to better address issues commonly encountered during AUV operation, such as the effects caused by ocean currents, a velocity synthesis method is adopted to navigate the vehicle in favor of ocean currents to optimize energy consumption. Zhu et al. [61] further studied the obstacle avoidance and speed jump problems of AUVs during task assignments when using SOM. These problems are addressed by integrating a new biologically inspired neural network, GBNN, to help AUVs avoid collisions with obstacles during underwater travel.

3.2.4 Heterogeneous AUV Task Allocation

To better complete a certain overall task, a multi-AUV system needs to have a variety of individuals with different functions, the heterogeneity of AUVs must be considered in the proposed methods [62]. Due to the heterogeneity between AUVs, heterogeneous AUVs exhibit stronger multi-task coupling and job conflict when working cooperatively, the impact of the functional structure of the AUV system on the task allocation of the multi-AUV system is needed to be taken into consideration, which makes it challenging to solve the task allocation of heterogeneous AUVs. How to give full play to the heterogeneity of different AUVs and effectively avoid task conflicts to achieve the overall optimal configuration is a research topic [63–65].

The above algorithms can well solve the task allocation of heterogeneous AUV systems. The difference from the homogeneous AUVs is that in terms of task allocation in a heterogeneous system, the task options performed by a certain AUV should be fully considered. In addition, the communication between AUVs must also be taken into account [64].

3.2.5 SOM Algorithm

Originally proposed by Kohonen [66], the SOM algorithm is an artificial neural network that uses unsupervised learning to generate a low-dimensional representation of the input space. The basic structure of SOM is composed of an input layer and an output layer, as shown in the Figure 3.5. The neurons to be trained are contained in the output layer, and the training process is realized by connecting the output layer to the input layer. The data input in the input layer is compared and analyzed with the competition layer, and the clustering effect is achieved by continuously updating

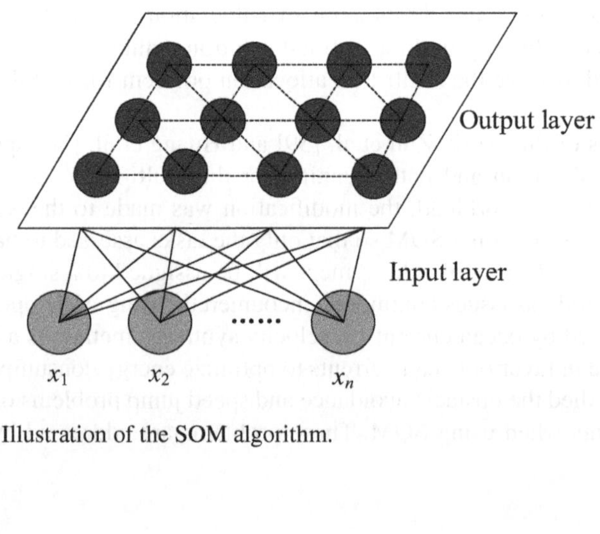

FIGURE 3.5 Illustration of the SOM algorithm.

the weights. When applying SOM to the task allocation, the input layer selects coordinates of the current target, while the value of the competition layer is set to AUV point coordinates.

Usually, the SOM algorithm is mainly divided into two steps:

Acquisition of winning neurons: Using A_{min} to represent the AUV closest to the target T_i, then the distance between the two can be expressed as:

$$D_{ij} = |T_i - A_j| = \sqrt{(T_{ix} - A_{jx})^2 + (T_{iy} - A_{jy})^2} \tag{3.4}$$

where D_{ij} is the Euclidean distance between j_{th} output neuron and i_{th} input neuron.

$$A_{win} = A_{min} = argmin\{D_{ij}\}, for\ j = 1,2,....,n \tag{3.5}$$

Among them, n represents the number of AUVs in the multi-AUV system.

Update and determination of neighborhood function (including weight update): After confirming the winning neuron, the neighborhood function needs to be updated. The purpose of this function is to spread the influence of the target to the remaining neurons in the neighborhood of the winning neuron. The mathematical expression of the neighborhood function is as follows:

$$f(d_j, g) = \begin{cases} e^{-dj^2/u^2(t)}, & d_j \leq D_{min} \\ 0, & d_j > D_{min} \end{cases} \tag{3.6}$$

$$u(t) = (1-\mu)^t g_o \tag{3.7}$$

Among them, D_{min} represents the neighborhood radius, t is the number of iterations, and μ represents decay rate, which is the time required to determine each iteration, which is inversely proportional to the time. By setting the values of g_0 and μ, the motion step length of the AUV can be changed to achieve the function of adjusting the accuracy.

The neighborhood function should be decided according to the actual situation. In the encircling problem in a multi-AUV system, to complete the task efficiently, D_{min} should choose an appropriate constant.

To make the algorithm converge, it is necessary to update the weights, that is, to make the AUVs in the multi-AUV system close to the target point.

$$R_{imin}(t+1) = \begin{cases} R_{imin} + \phi(t) f(d_{imin}, g)(T_i - R_{imin}) & D_{imin} > D_{min} \\ T_i & D_{imin} \leq D_{min} \end{cases} \tag{3.8}$$

Among them, R_{imin} represents the position coordinates of the winning point corresponding to the completion rate of the i_{th} objective, and T_i represents the coordinates of the target point. The function $\phi(t)$ represents the learning rate, and D_{min} represents the shortest distance for a single AUV to reach the target point.

3.2.6 Improved SOM Algorithm to Form an Efficient Encircling Alliance

When using the traditional SOM algorithm to deal with task allocation, due to the energy limit of AUV, the load balancing in this problem should be fully considered. To realize the load balancing in the task allocation process, a load balancing adjustment function needs to be set for the SOM algorithm competition. This function can effectively combine the AUV load problem with the SOM competition principle.

3.2.6.1 Estimated Encircling Time

Suppose n AUVs should round up m invading robots. The positions of the invading robots and AUVs are represented by matrices E and W, respectively.

$$E = \begin{bmatrix} x_1 & y_1 & z_1 \\ \vdots & \vdots & \vdots \\ x_m & y_m & z_m \end{bmatrix} \tag{3.9}$$

$$W = \begin{bmatrix} w_{1x} & w_{1y} & w_{1z} \\ \vdots & \vdots & \vdots \\ w_{nx} & w_{ny} & w_{nz} \end{bmatrix} \tag{3.10}$$

D_{we} is the distance matrix formed by the distance of each AUV to all the invading robots.

$$D_{we} = \sqrt{W^2 \cdot M_1 + M_2 \cdot (E')^2 - 2 \cdot W \cdot E'} \tag{3.11}$$

where M_1 is a matrix with 3 rows and m columns, and M_2 is a matrix with n rows and 3 columns. All elements in the matrices M_1 and M_2 are 1s. E means the transpose of matrix E. The symbols W_2 and $(E')^2$ mean to square all elements of the matrices W and E. Therefore, the ith row vector in D_{we} represents the distance from the i_{th} AUV to all invading robots numbered from 1 to m.

For underwater currents and obstacles, in the case of currents, use formula to correct the distance cost, and when the AUV encounters obstacles during encircling process, use D*-lite re-planning to calculate the encircling distance.

Heterogeneous AUVs have different batteries and thrusters. These differences are mainly reflected in the safe distance and the maximum speed of AUV. Here, the distance that the AUV can run very safely without worrying about running out of energy is called safety distance. Considering the safety distance, if the total sailing distance of the AUV chasing an intruding robot will exceed the safety distance, the distance from the AUV to each intruding robot will be set to ∞.

$$d_{wiej} = \begin{cases} d_{wiej} & if\ d_{wiei} \le c_{safei} \\ \infty & ,otherwise \end{cases} \tag{3.12}$$

d_{wiej} is the distance from the i_{th} AUV to the j_{th} invading robot, it is the element value of the i_{th} row, j_{th} column of the matrix D_{we}, c_{safei} is the safe distance of the i_{th} AUV, $i = 1, 2\ldots n, j = 1,2\ldots m$.

In order to obtain the estimated encircling time, the speed of each single AUV is divided by the distance matrix from the AUV to the invading robot according to equation (3.13), that is, each row of D_{we} is divided by the speed of the corresponding AUV. Then each row of t_{we} represents the time, which takes for each AUV to hunt the invading robot.

$$t_{we} = \frac{(D_{we}).}{Vel} \tag{3.13}$$

where V_{el} is the velocity matrix containing the velocity values of every single AUV, and edited to the same number of rows and columns as matrix D_{we}, each row of the velocity matrix V_{el} is the velocity of the i_{th} AUV.

3.2.6.2 Time-based Competition Mechanism to Generate Encircling Coalitions

For a certain intruding robot, the estimated encircle time can be obtained by the matrix t_{we}. First, the invading bots schedule a round-up sequence by the estimated encircling time. The AUVs are then grouped to encircle the invading robots one by one. A summary of the competition mechanism is shown in Algorithm 1.

In the heterogeneous multi-AUV cooperative encircling problem, AUVs are usually of different types and have different capabilities. In the early stage, researchers proposed that the encircle alliance generation algorithm usually only considers the distance from the AUV to the invading robot, and does not consider the ability difference of each AUV. Therefore, the constructed coalition may lead to inefficiencies in encircling, thus requiring more time. In this paper, a new encircle alliance generation algorithm based on the time competition mechanism is adopted, and the different speeds of AUVs are considered so that the formed encircle alliance can realize tasks with lower energy costs and less time. In addition, in order to avoid energy exhaustion caused by the low energy configuration of some AUVs, the algorithm also considers the safe distance of AUVs.

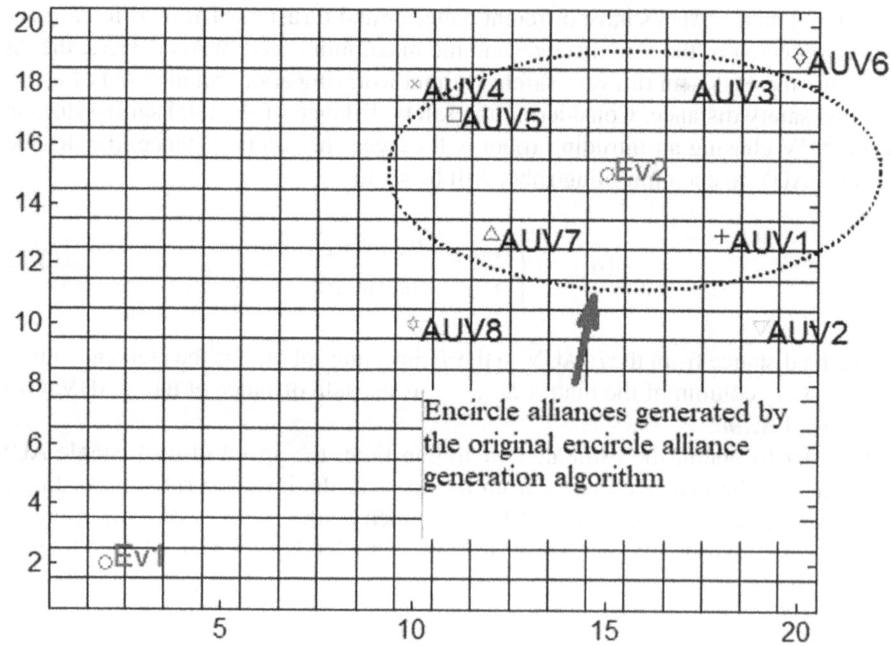

FIGURE 3.6 An illustration of the merits of forming an encircling coalition.

Algorithm 1: Time-based competition mechanism to generate encircling coalitions

Input: t_{we}: Estimated encircling time, TeamNum: The maximum number of teams an AUV can form, Num: Number of AUVs required per encircle team, Encircling(2-D): Num = 4, Encircling(3-D): Num = 6; NumEvader: The number of invading robots
Output: evaderIndex: Encircling order for invading bots (row vector), AUVIndex: Encircle alliance formed by AUVs corresponding to the encircling sequence of invading robots
Initialization: j = 1, evaderIndex = 1 row TeamNum column zero vector, AUVIndex = Matrix of zeros with TeamNum rows and Num columns
While j <= NumEvader && j <= TeamNum
Sort the encircle time and return the sorted time and corresponding AUV sequence
For i=1 to NumEvader
Get the first Num rows of sortedTime
End For
Get the serial number of the intruding robot with the smallest encircling time
Assign AUVs to form a team to encircle the invading robot
Set the encircling time of the intruding robot to be encircled to a larger value to prevent it from being assigned again
j = j+1
End While

According to the estimated encircling time, the most easily captured robot can be determined, and the most easily captured intruding robot will be selected as the first encircled target of the AUV. In response to this intruding robot, the AUV with the shortest encircling time is expected to form an alliance team and start the hunt. During the chase, the invading robot will try to escape. This will cause the AUV team with the least hunting time to change, and thus the alliance, so the generated encircling alliance is dynamically changed according to the corresponding rounding environmental conditions.

Figure 3.6 explains the process of forming an encircling coalition and illustrates the advantages of the proposed encircling coalition generation algorithm. In Figure 3.6, it is assumed that AUV1, AUV3, AUV5, and AUV7 are AUVs with a maximum speed of 1. While other AUVs can reach twice the speed of AUV1, AUV3, AUV5, and AUV7. The traditional alliance generation method depends on the distance between the AUV and the hunting point. The formed encircling alliance includes the AUVs in the dotted ellipse in Figure 3.6, that is, AUV1, AUV3, AUV5, and AUV7. If the invading robot tries to escape, then when all the AUVs and the invading robot have the same speed, the insufficient speed of the AUV may not immediately stop the target from escaping, causing the target to escape and making the hunting fail. With the proposed time competition mechanism, different alliances are formed. AUVs with faster speed will win the task of encircling Ev2, such as AUV4 and AUV6 will replace AUV5 and AUV3, forming a hunting team with AUV1 and AUV7 to encircle Ev2, while the rest of the AUVs will start to encircle Ev1 first. Ev2 is referred to here as the easiest intrusion robot to capture and will be encircled first, and the encircle team will be formed in the AUV with the shortest time, which is beneficial to quickly capture the intelligent intrusion robot.

3.3 PURSUIT PATH PLANNING IN MULTI-AUV COLLABORATIVE ENCIRCLING

In the field of robotics research, correct and rational path planning is an extremely critical part of the process to ensure the efficiency and safety of its motion. Usually, path planning refers to the selection of the most efficient motion path in a complex working environment under the premise of guaranteeing safety according to certain evaluation criteria. In robot path planning technology, the most critical part of two factors, one is obstacle avoidance, and one is path optimization. Obstacle avoidance guarantees its safety, and path optimization ensures its efficiency. The robot senses itself and its environment through sensors and uses different computational methods to achieve the optimal path from the start point to the end target. A reasonable path enables the robot to work efficiently and safely, while an unreasonable path will not only increase energy consumption and cost but also increase the probability of accidents. Therefore, an in-depth study of path planning is still necessary.

3.3.1 A* Algorithm

The A* algorithm is widely adopted in global path planning, which uses heuristic search and combines with the breadth-first algorithm to find the optimal solution in a static environment. The A* algorithm is formed by combining the conventional method Dijkstra's algorithm with the best-first search algorithm. By defining a cost function:

$$f(x) = g(x) + h(x) \tag{3.14}$$

In the formula x represents the current node, $f(x)$ is the cost function of the current node, $g(x)$ is the actual generation value of the starting point to the current node, and $h(x)$ is the generation value from the current node x to the target.

The A* algorithm continuously traverses the neighboring points around the starting point in the direction of the target point, calculates the generation value of each surrounding node by the heuristic function $h(x)$, selects the minimum generation value as the next expansion point, iterates the cycle until the search stops at the target point, then returns to the starting point, and plans the optimal path according to the cost function. The Euclidean distance is chosen as the cost function of $h(x)$ in the case that the robot moves in four directions, the function can be expressed as:

$$h(x) = \sqrt{\left(x_n - x_g\right)^2 + \left(y_n - y_g\right)^2} \tag{3.15}$$

In the equation, (x_n, y_n) represents the grid center coordinates where the current path node is located, and (x_g, y_g) represents the grid center coordinates where the target node is located.

The traditional A* algorithm enables the mobile robot to find the optimal path accurately, but its planned path often has disadvantages such as redundant nodes, too many inflection points, more path folds, and too large turning angles at the path nodes without conforming to the kinematics of the mobile robot. José Lima [30] proposed a fusion algorithm based on the combination of the improved A* algorithm and dynamic window strategy by improving this way for key point selection, fusing the dynamic window method, constructing the globally optimal path evaluation function, and using ultrasonic sensors for obstacle avoidance, which ensures the smoothness and real-time performance of the path. Maneev [67] used the algorithm to extend the sub-node jumping based on the traditional A* algorithm by diffusing the sub-node jumping method instead of the A* algorithm to calculate the surrounding raster. Thus, the pathfinding efficiency is improved and the smoothness optimization of the Bessel curve is used to improve the driving stability. Chih [68] proposed an improved A* algorithm, which introduces a time factor to reduce the number of turns and effectively avoids the conflict and deadlock problems of the system by combining the global dynamic planning with the time window model and the priority strategy.

3.3.2 Fuzzy Logic Algorithm

Fuzzy logic control algorithms combine the robustness of fuzzy control with physiologically based "perception-action" behavior to provide new ideas for obstacle avoidance and navigation of mobile agents in complex environments. The fuzzy logic path planning algorithm uses ultrasonic sensors to collect external information, and a fuzzy controller to process the collected information and control speed and direction of the robots according to the processed structure, thus achieving real-time obstacle avoidance. The fuzzy control algorithm is able to perform path planning in extremely complex environments, with a weak dependence on environmental information and high adaptability. And controlling the speed makes the robot save time than the previous path planning with only corner control, which has strong timeliness. Bakdi [69] combined the principle of line-of-sight navigation with a fuzzy control algorithm to propose a local path planning strategy for AUV in an unknown underwater environment, which can realize the real-time obstacle avoidance function of the robot. Jiang [70] proposed a fuzzy obstacle avoidance algorithm with velocity feedback, which improves the stability and effectiveness of environmental information through multi-sensor data fusion processing by grouping obstacle information and setting weights.

3.3.3 Simulated Annealing Algorithm

The simulated annealing algorithm mimics the principle of annealing in industry and proposes an optimization search method that mimics the annealing process. By imitating the process of material annealing, the initial temperature, initial state, and cooling rate are set to control the continuous decrease of temperature, and the probabilistic sudden jump property is combined with the neighborhood structure of the solution space for a random search. The simulated annealing algorithm does not depend on the initial solution, has a simple structure, is more efficient, and can go out of the local minimal value point. However, the stochasticity is strong and the convergence speed is slow.

The grid method was combined with the simulated annealing (SA) method [71], which was used to represent the environment information. And the local path planning was mainly based on the simulated annealing method to make the path jump out of the local minima, and eventually, reach the target location. Capraz [72] combined the GA and SA algorithms, then, proposed a hybrid GA-SA computation method, which effectively improved the convergence speed. Behnck [73] proposed a multi-objective annealing algorithm (MOSA) and implemented a perturbation to select a new solution based on the MOSA algorithm, which has better robustness. Fang et al. [74] combined the artificial potential field method with the simulated annealing algorithm to implement obstacle avoidance by adding the influence distance limit value to the gravitational field function and introducing the scheme of the simulated annealing method, which has a positive effect on real-time obstacle avoidance. The author [75] proposed a method to fuse ACO and SA algorithms for PID controller parameter optimization, replacing the ACO algorithm with the search method of SA to avoid the algorithm falling into local solutions, and proposed a hybrid optimization algorithm based on the SA and PSO algorithms to avoid the disadvantage of probabilistic abruptness in the pure simulated annealing method.

3.3.4 Neural Network Method

Neural network, as a significant part of artificial intelligence, has received extensive attention in the study of mobile robot path planning. It performs distributed parallel information processing by simulating the behavior of animal neural networks. The neural network has excellent learning ability, but the changing underwater environment in path planning is tough to describe by mathematical formulas, which leads to its poor generalization ability. Due to its better learning ability and robustness, the application, which combines with other algorithms, is a major trend in the current path planning of neural network strategy.

Mokhtari [76] proposed an improved neural for ship navigation path planning, selecting a modified linear unit ReLU function to replace the original activation function, updating the neural network weights with the gradient descent method, and optimizing the neural network using the root mean square backpropagation algorithm to achieve effective planning of motion paths and avoid collisions. Sung et al. [77] adopted a method based on an adaptive fuzzy neural network, which integrates the autonomous learning function of the neural network with the fuzzy reasoning ability, and proposed a navigation controller based on an adaptive fuzzy neural network, which outputs the perturbation angle in real-time and adjusts the pre-targeting direction of the mobile robot online so that the mobile robot can converge to the target without collision, and proposed an improved virtual target method that prioritizes the path where the robot may escape the trap state, simplifies the design difficulty, changes the virtual target switching, and avoids a large number of complex calculations. Immanuel Johanes [78] proposed a bio-inspired neural network algorithm based on path correction and obstacle-free ideal path guidance, which solved the path deviation problem in the initial phase, then made the cost and the number of path turns lower and more efficient.

In recent years, artificial neural network algorithms have been widely used in various fields, but because of their drawbacks such as cumbersome learning time and learning lag, there is still room for deeper research on artificial neural network algorithms in robot path planning. In this regard, Simon Yang, a Canadian scholar, proposed a non-learning, adaptive GBNN and applied it to full-coverage path planning for mobile agents. The basic principle of this algorithm is to build a GBNN model to represent the working environment of AUV based on the raster map so that the raster map corresponds to the neural network one by one. Based on the distribution of the active output values of each neuron, the AUVs are able to run in a collision-free and energy-saving navigation path autonomously.

The basic idea of the GBNN algorithm is that the target continuously transmits the excitation outward, while the obstacles have an inhibitory effect on the excitation, and the activity value of each location is calculated by iteration. The mathematical description of the GBNN model is as follows:

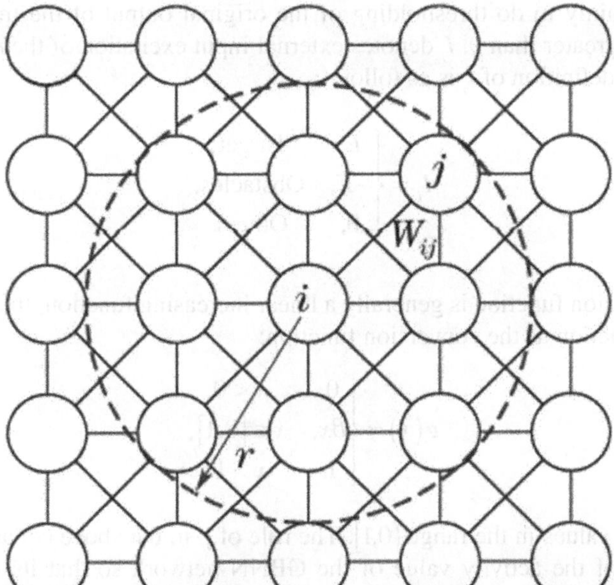

FIGURE 3.7 GBNN model diagram.

$$x_i(t+1) = g\left(\sum_{j=1}^{M} W_{ij}\left[x_j(t)\right]^+ + I_i\right) \tag{3.16}$$

$$w_{ij} = \begin{cases} e^{-y|i-j|^2}, & |i-j| \leqslant r, \\ 0, & |i-j| > r, \end{cases} \tag{3.17}$$

$$|i-j| = \sqrt{\left(x_{ix} - x_{jx}\right)^2 + \left(x_{iy} - x_{jy}\right)^2} \tag{3.18}$$

$$|i-j| = \sqrt{\left(x_{ix} - x_{jxx}\right)^2 + \left(x_{iy} - x_{jy}\right)^2 + \left(x_{iz} - x_{jz}\right)^2} \tag{3.19}$$

the output activity value of neuron i, W is the connection coefficient matrix, its network structure is shown in Figure 3.7. From its network structure, it is easy to know that the connection coefficient matrix W is a symmetric matrix; w_{ij} represents the connection coefficient between the i_{th} neuron and the j_{th} neuron.

$|i$-$j|$ denotes the Euclidean distance between the i_{th} neuron and the j_{th} neuron, where equation (3.18) is the Euclidean distance of the corresponding GBNN model in two dimensions, and equation (3.19) is the Euclidean distance of the corresponding GBNN model in three dimensions. The function g is the transformation function of the model, mainly to do thresholding of the original output of the model. γ and r are constants greater than 0. I_i denotes external input excitation of the i_{th} neuron, the mathematical definition of I_i is as follows:

$$I_i = \begin{cases} E, & \text{Target,} \\ -E, & \text{Obstacles,} \\ 0, & \text{Others.} \end{cases} \tag{3.20}$$

The conversion function is generally a linear increasing function, this paper uses a segmented function as the conversion function:

$$g(x) = \begin{cases} 0, & x < 0, \\ \beta x, & x \in [0,1], \\ 1, & x > 1. \end{cases} \tag{3.21}$$

where β takes values in the range $(0,1]$. The role of β in the above equation is to perform scaling of the activity value of the GBNN network so that its orientation is limited between 0 and 1. Although the algorithm works well as long as the value of β satisfies the condition of $(0,1]$. However, choosing a small value of β can cause the GBNN activity to decay too fast, especially at points far from the target. Its activity value can usually reach more than the negative 50th power of 10. In a GBNN model, the activity value should be less than 1 at all locations except at the target, where the activity value is 1. Due to the updating characteristics of the network, the activity value has a cumulative superposition, so it is necessary to ensure that the activity value at the non-target is less than 1, otherwise, the neural network activity value cannot be transferred properly, which is the basic condition to be satisfied in the GBNN model. The parameters γ and in (3-3) ~ (3-5) jointly determine the magnitude of the activity value of the neurons in the network. In path planning, the path is selected by comparing the size of the activity value of the GBNN network, so a high precision processor is required, so in the actual path planning problem, the β value is usually taken as 1 by choosing the appropriate γ value, which has the advantage of preventing activity value from decaying too fast in the GBNN.

3.3.5 Path Planning Algorithm in Multi-AUV Collaborative Encircling

This section introduces two different pursuit path planning algorithms in encircling of multi-AUV system. After simulation analysis, both of them can quickly plan a reliable and collision-free path for AUVs to encircle the intruding robot.

3.3.5.1 Pursuit Path Planning Algorithm Based on Integrated Belief Function and Original GBNN Neural Network

First of all, we discuss the path planning algorithm combining the reliability function path planning and the original GBNN neural network for the encircling AUV to complete the hunt of the intruding robot. The path planning algorithm of the reliability function and the original GBNN neural network will not be repeated here. Combining the two algorithms, the comprehensive calculation formula of neuron activity is shown in formula (3-9).

$$y_i(t+1) = x_i(t+1) + e^{-(m+t)} \cdot b(i) \tag{3.22}$$

Equation (3.22), y_i is the total neuronal activity outside the current position of the AUV. P_p, P_c, and P_n are the previous, current, and next time step positions of the AUV, that is, according to the equation, the AUV will sail to the grid with the greatest total neuron activity among the grids adjacent to it.

Equation (3.22) adds the neural activity of the GBNN to the reliability function value and uses this comprehensive activity value to plan the path of the AUV. m is constant and set to a large initial value, and the time term in the exponential function e^(-(m+t)) can make an effect on the belief function value decrease with time. As with a single GBNN or belief function algorithm, the AUV will navigate to the grid adjacent to it with maximum total neuron activity.

The integrated belief function and the GBNN neural network algorithm applies the belief function as a local effect, when the AUV is far from its target, the activity value of the belief function is relatively large due to the small GBNN activity value in this area. As time goes by, the AUV gradually approaches the target, and the activity value of GBNN increases, which will be responsible for guiding the AUV to reach the round-up occupancy point. The neuron activity of GBNN in the obstacle unit can be equal to zero, and the neuron activity of the belief function algorithm is less than zero. However, the combined neuronal activity of idle grid cells in the direction close to the target is greater than zero and increases as the AUV's distance from the target decreases. Therefore, even if the target or obstacle is dynamic, the comprehensive path planning algorithm will ensure that the neuron activity of the obstacle is smaller than that of the free unit and gradually increases along the target direction. That is, neuronal activity propagates from the target to the AUV and has locally smaller neuronal activity values at the obstacle point. In a word, AUV takes the grid with the maximum total neuron activity as the direction and position of its next time step, which can complete obstacle avoidance and finally reach the target position safely.

All in all, the propagation of neuron activity in the original GBNN model is somewhat like the propagation of water waves in a lake. If the waves start from the center of the lake, they will take some time to spread to the shore. The wider the lake, the longer the spread. A single GBNN model takes some time to spread the neuron activity at the target grid to the location of the AUV. The integrated algorithm applies the belief function as a local effect, and when the AUV is far from its target, this local effect is relatively large due to the small activity value at that location. The comprehensive algorithm can clear up the time delay of GBNN model breeding, which is beneficial for AUV to move quickly to hunt down intruding robots.

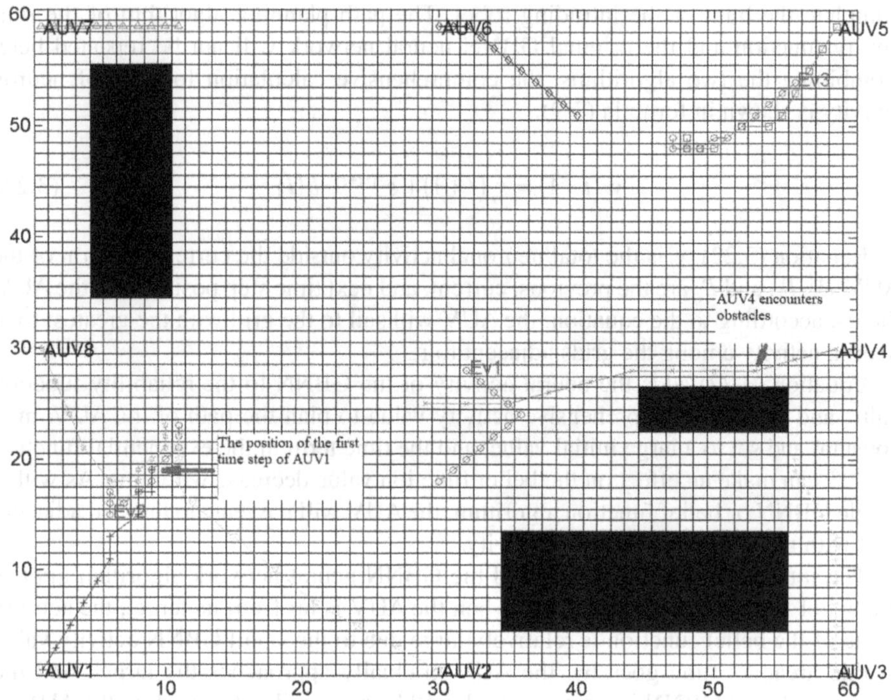

FIGURE 3.8 The encircling process in an underwater environment(including obstacles) of the integrated path planning algorithm (the 30th time step).

3.3.5.2 Pursuit Path Planning Algorithm Based on Improved GBNN Model with Inverse-proportional Weights

Then, this chapter will also discuss the path planning algorithm of the encircling AUV using the improved inverse-proportional GBNN neural network to complete the pursuit of the intruding robot.

For the purpose of facilitating the simulation analysis, three reasonable assumptions are made. Firstly, due to the vast ocean environment, the shapes of the invading robots and AUVs are ignored, that is, the intruding robots and AUVs are treated as particle points. Secondly, it is assumed that the AUV can turn slowly, following the path planned by the raster map-based path planning algorithm. Since AUVs are generally slower, this assumption is reasonable. And the AUV has plenty of time to change its direction to follow the planned path. Finally, after one of the AUVs finds the intruding robot through the hydroacoustic sensor network, it is assumed that all the AUVs can quickly know the location information of the intruding robot. The location information of the intruding robot is a short message, which can be quickly transmitted and notified to all AUVs through underwater acoustic communication.

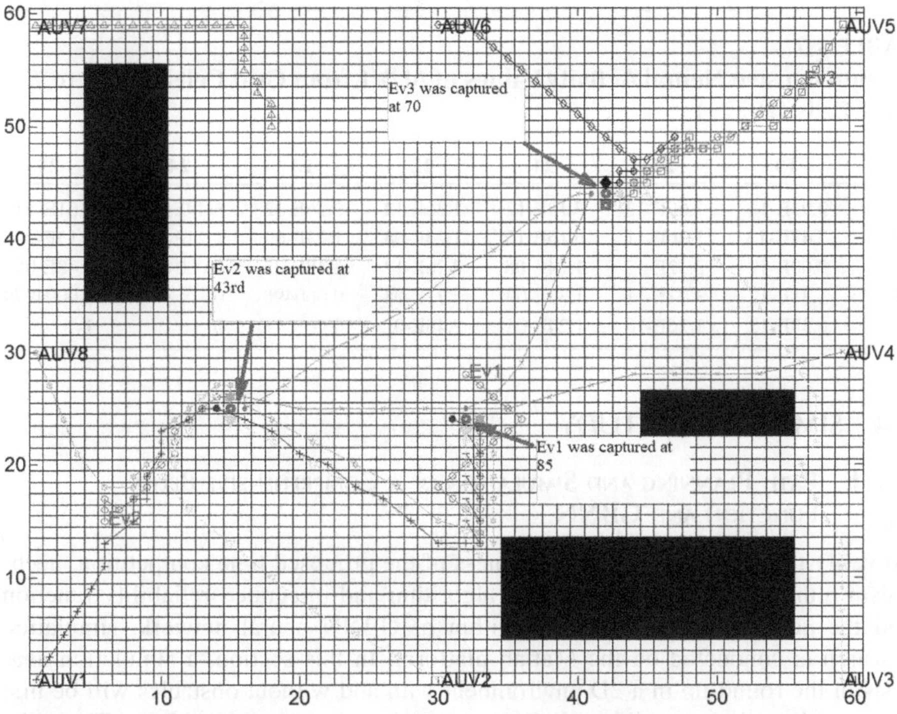

FIGURE 3.9 The encircling process of the integrated path planning algorithm in an underwater environment (including obstacles) (encircling completed).

TABLE 3.1
Comprehensive Neural Activity Values of AUV4 from Grid Cell (53, 28) to Grid Cell (50, 28)

(x, y)	25	26	27	28	29	30	31
50	-9.73E-08	-9.64E-08	-9.58E-08	**1.63E-08**	1.48E-08	1.28E-08	1.10E-08
51	-9.87E-08	-9.74E-08	-9.61E-08	1.62E-08	1.38E-08	1.09E-08	8.89E-09
52	-1.01E-07	-9.98E-08	-9.75E-08	1.61E-08	1.09E-08	7.48E-09	5.87E-09
53	-1.04E-07	-1.04E-07	-1.04E-07	**5.18E-09**	2.26E-09	2.23E-09	2.20E-09
54	-1.08E-07	-1.10E-07	-1.13E-07	-5.78E-09	-4.67E-09	-2.41E-09	-1.17E-09
55	-1.11E-07	-1.13E-07	-1.16E-07	-5.86E-09	-6.03E-09	-4.77E-09	-3.50E-09
56	-1.13E-07	-1.15E-07	-1.17E-07	-5.94E-09	-6.32E-09	-5.76E-09	-4.87E-09

TABLE 3.2
Comprehensive Neural Activity Values of AUV1 from Grid Cells (9, 19) to Grid Cells (20, 21)

(x, y)	19	20	21	22	23	24	25
8	3.91E-13	7.60E-13	7.12E-12	1.36E-12	6.56E-12	6.49E-13	1.61E-14
9	**7.47E-13**	1.36E-11	6.16E-10	1.24E-08	1.23E-09	1.24E-08	6.15E-10
10	7.00E-12	2.32E-12	**1.24E-08**	2.26E-12	2.48E-05	0.0004978	2.48E-05
11	1.31E-12	3.77E-11	1.35E-09	3.71E-08	0.000498	0	0.0004979
12	6.38E-12	1.26E-10	1.24E-08	2.48E-07	0	0	**1**

3.4 SIMULATION STUDY

3.4.1 Path Planning and Simulation of a Comprehensive Belief Function and GBNN

To verify the feasibility and effectiveness of the proposed time competition mechanism round-up coalition generation algorithm and integrated reliability function and the pursuit path planning algorithm of GBNN neural network, simulation research is conducted on the Matlab platform. In this section, a simulation analysis of the round-up in a 2D environment with and without obstacles will be first performed, and the results will be compared with other algorithms. Then, the cooperative encircling simulation of multi-AUV in a 3-D underwater environment is completed.

AUVs and intelligent intrusion robots are regarded as particles in this paper. In this section, the obstacles will be enlarged by one grid cell size to consider their actual shape. However, if the enlarged obstacle grid is the occupation point of the intrusion robot, the AUV can also enter the occupation point, while the intelligent intrusion robot is not allowed to enter the enlarged obstacle space.

3.4.1.1 Simulation Design in 2-D Environment

In the simulation, 8 AUVs will encircle 3 intelligent intrusion robots. The invading robots are named Ev1, Ev2, and Ev3, and the AUVs are named AUV1, AUV2, …, and AUV8. The simulated environment is a 2-D area with a 60×60 grid. The simulation environment is displayed with a scale of 1:100, and the size of the underwater environment is 6*6 km. It is assumed that the perception circle of the intelligent intrusion robot to encircle the AUV is a circular area with a diameter of 10 grids. When AUV enters the perception circle of the intelligent intrusion robot, the intelligent intrusion robot detects the AUV and will try to escape. Here it is assumed that the speed of the intelligent intrusion robot is 1. AUV1, AUV3, and AUV5 have a speed of 2, while AUV2, AUV4, and AUV8 have a speed of 3. The other two AUVs have the same speed as the intelligent intrusion robot. Therefore, the intelligent intrusion robot can move one grid per time step, but the faster AUV can move to go to grids per time step. And the safe sailing distance of AUVs with a set speed of 1 is 12,000 meters, the safe

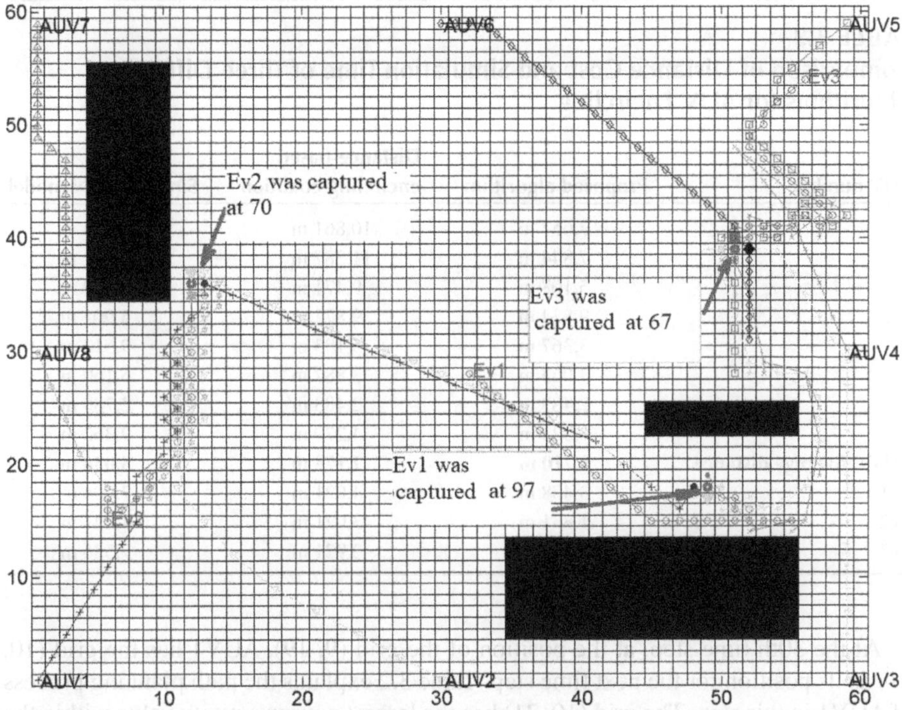

FIGURE 3.10 Simulation results of the coalition generation algorithm based on distance competition.

sailing distance of AUVs with a speed of 2 is 18,000 meters, and the energy of the fastest AUVs can make their safe sailing distances reach 22,000 meters.

This paper mainly conducts the simulation analysis of encircling in a two-dimensional environment with obstacles. Except for the presence of obstacles in the environment, other environmental conditions (including parameter settings) are unchanged. The round-up process is shown in Figures 3.8 and 3.9. The black areas in the figure represent the obstacles.

Figure 3.8 shows the AUV encircling process and states for the first 30-time steps. In the initial stage of the round-up, the AUVs formed two encircling teams to encircle Ev2 and Ev3. When encountering an encircling AUV, the intelligent intrusion robot tries to change direction to escape. AUV4 encountered the obstacle at the grid (53, 28). Table 3.1 shows the comprehensive neural activity values of AUV4 around the grid (53, 28). The first three columns in the table are the neuron activity values of grid cells with obstacles, and it can be seen that the total neuron activity values of these grid cells are all much less than 0. AUV4 can move three grid cells in one time step. Here, it finds the largest neuron activity value at the position (50, 28) within three grids from the current position, so it moves to (50, 28). Therefore, AUV4 can avoid obstacles and navigate safely through this comprehensive path planning algorithm.

TABLE 3.3
Comparison of Distance Cost and Simulation Time of Three Different Algorithms for AUV Encircling

AUV number	Proposed algorithm	Distance-based encircling coalition	Single GBNN model
1	7,657 m	10,861 m	8,364 m
2	7,844 m	11,387 m	10,431 m
3	5,108 m	9,471 m	5,929 m
4	9,834 m	9,872 m	5,901 m
5	3,367 m	5,193 m	9,512 m
6	3,063 m	3,887 m	6,160 m
7	2,483 m	2,483 m	2,780 m
8	6,323 m	4,270 m	9,362 m
AUVs' Average distance	**5,710 m**	**7,178 m**	**8,034 m**
Ev1	3,438 m	4,021 m	7,004 m
Ev2	1,831 m	3,021 m	4,221 m
Ev3	2,673 m	2,921 m	7,536 m

At the 30th time step, at the position of the grid (9, 19), AUV1 has the grid (10, 21) as its position for the next time step. Table 3.2 explains the path planning process of AUV1 in this step. The grid (10, 21) has the largest neuron activity value within the grid range that AUV1 can reach in one time step. As can be seen from Table 3.2, the encircling occupancy points assigned to AUV1, as the target points of AUV1, have a neuron activity of 1. The grid cells occupied by other AUVs and invading robots are regarded as obstacles for AUV1, and their neuron activity is less than or equal to zero, so, AUV1 will not enter the location of these grids, which can avoid collision with other AUVs and invading robots. In conclusion, guided by neuronal activity values, AUV1 can safely gradually approach the encircling occupancy site to capture Ev2.

Ev1, Ev2 and Ev3 are captured at grid positions (32, 24), (15, 25) and (42, 44), respectively. Ev2 was firstly encircled by AUV1, AUV2, AUV4, and AUV8. After capturing Ev2, multiple AUVs re-formed an alliance. At this point, AUV3, AUV4, AUV5, and AUV6 were allocated to encircle Ev3, and the remaining AUVs encircled Ev1. Then, an encircling team consisting of AUV3, AUV4, AUV5, and AUV6 captured Ev3 at the 70th time step. Finally, AUV1, AUV2, AUV4, and AUV8 form the encircling alliance and start hunting Ev1 until the round-up mission is successfully completed. The second column of Table 3.3 lists the total distance cost of the AUV's tracking and the escape distance of the intruding robot.

3.4.1.2 Comparative Simulation in 2-D Environment

This section will conduct a comparative simulation analysis of the multi-AUV cooperative round-up. First, a comparative simulation of different encircling coalition generation algorithms is adopted. The results show that the dynamic encircling alliance strategy based on encircling time proposed in this chapter is more effective and superior to the original distance alliance generation algorithm, which enhances

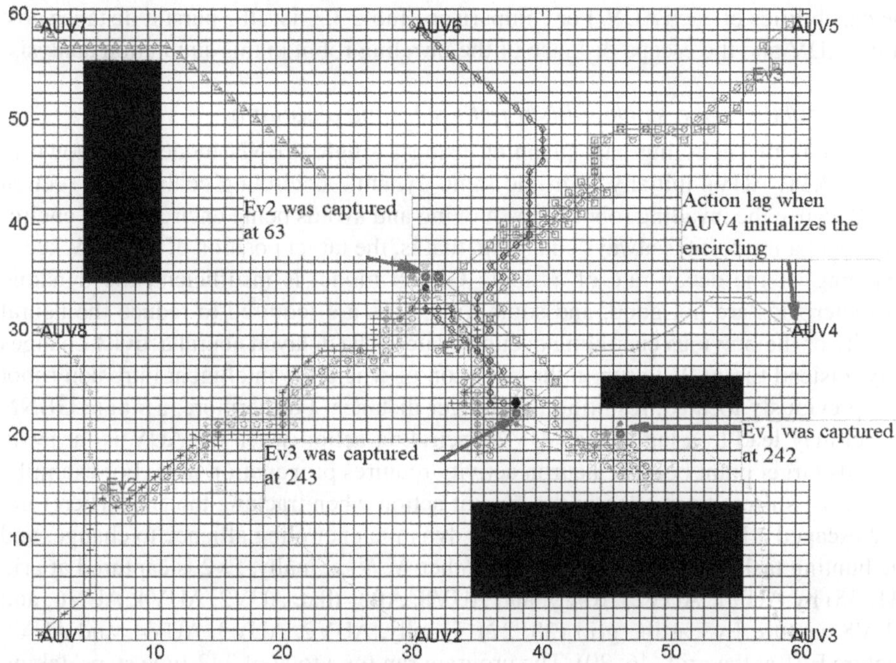

FIGURE 3.11 Simulation results of multi-AUV encircling for a single GBNN algorithm.

the ability of the encircling alliance and improves the success of the encircling rate. Then, the comprehensive pursuit path planning strategy is compared with that of the single original GBNN model, and the result proves that the encircling and tracking of the comprehensive path planning algorithm is faster and more efficient.

(1) Simulation of the distance-based encircling coalition generation algorithm

Based on the same environment setting, a traditional dynamic distance-based alliance competition mechanism is applied. The results are shown in Figure 3.10. In this way, AUV forms dynamic encircling alliances in different ways. AUV1, AUV2, AUV7 and AUV8 round up Ev2 first. Another team contains other AUVs encircling the Ev3. The results show that the distance-based competition mechanism and the comprehensive path planning algorithm can also complete the encircling, but it takes more time and distance. The intruding units are hunted at grids (49, 18), (12, 36), and (51, 39), respectively. Ev3 was firstly hunted by AUV3, AUV4, AUV5, and AUV7 at time step 67. Then, at the 70th time step, AUV1, AUV2, and AUV8 capture Ev2 with the help of obstacles. Finally, AUV1, AUV2, AUV3, and AUV4 capture Ev1 at the 97th time step. Since the formation of the encircling alliance depends only on distance competition, some AUVs in the alliance have lower sailing speeds, allowing invading robots to escape longer distances. Therefore, the encircling alliance formed by distance competition is not suitable in a heterogeneous AUV system, because it ignores the capabilities of each AUV. For comparison, Table 3.3 lists the pursuit distance cost of the AUV and the escape distance of the invading robot in this simulation analysis.

(2) Simulation of a single GBNN path planning algorithm

The results of encircling path planning using a single GBNN model are shown in Figure 3.11. AUVs initially form the same dynamic encircling coalition as section 2.6. Because the invading robot keeps moving and avoids being encircled, the encircling occupancy point is always changing, that is, the target position of the AUV keeps changing. If the decay rate of a single GBNN model is insufficient, the dynamic characteristics are not good, and some time steps are required to reduce the neural activity of the previous encircling occupied sites. The delayed neural activity changes may mislead the AUV to run in the direction of where the intelligent intrusion robot was previously located, resulting in a longer distance. Furthermore, a single GBNN model may lead to a lag in the AUV's encircling action when the AUV is far away from its target point because neuron activity requires propagation time. For example, the AUV4 suffers from a lag in the pursuit action when tracking the Ev2. Therefore, Ev2 escaped a long distance, causing the dynamic encircling alliance to change, and the hunting task of Ev2 was assigned to other AUVs. Finally, Ev2 is captured at grid (31, 35) by AUV1, AUV2, AUV5, and AUV8. After that, AUV2, AUV4, AUV6, and AUV8 capture Ev3 at the grid (38, 22). Finally, AUV2, AUV3, AUV4, and AUV8 capture Ev1 at the grid (46, 20). The program ran for a total of 242-time steps, taking more encircling time and distance than the other two simulations above. The distance cost of AUV encircling and the escape distance of invading robots are listed in Table 3.3.

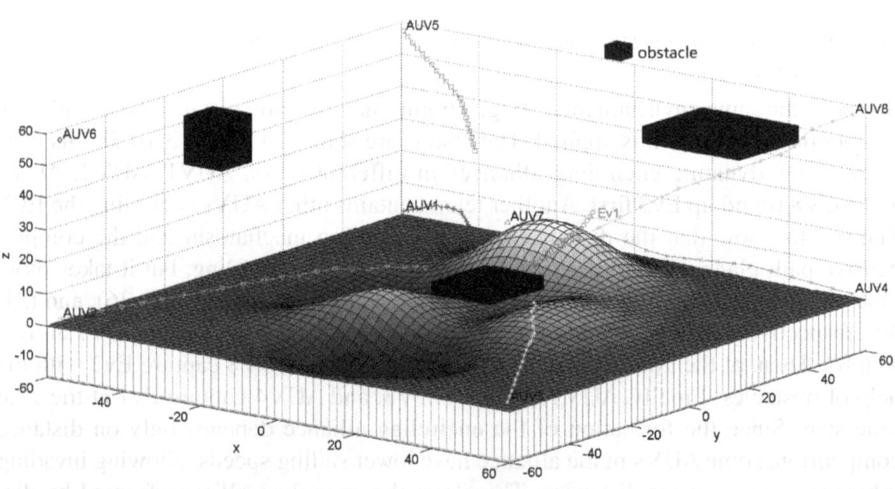

FIGURE 3.12 Encircling process of a 3-D underwater environment with time competition mechanism and integrated path planning (time step 20).

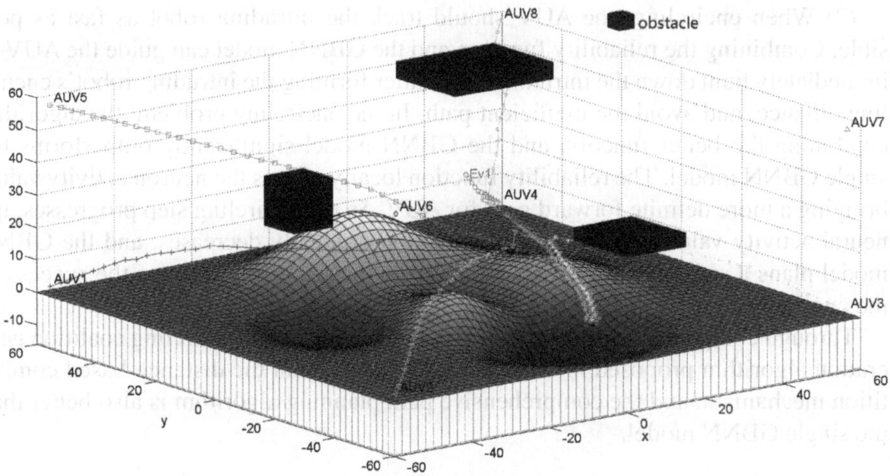

FIGURE 3.13 Encircling process of a 3-D underwater environment with time competition mechanism and comprehensive path planning (task completed).

3.4.1.3 Analysis of Simulation Results

The results of the above comparative simulation studies are listed in Table 3.3. The travel distance spent by every AUV in the encircling process and its average distance are listed in the table, and the escape distance of the invading robot is listed for the efficiency comparison of the three different algorithms.

From Table 3.3, the proposed algorithm can help AUVs capture intruding robots with less encircling distance cost and higher efficiency. Based on the proposed alliance algorithm and a comprehensive path planning algorithm, an average distance of 5710 meters is required for the AUV to successfully capture all invading robots. However, the distance competition-based encircling coalition generation and the single GBNN model require each AUV to travel an average of 7178 and 8034 meters to capture the intruding robot, respectively, and the round-up efficiency drops significantly.

(1) When heterogeneous AUVs hunt down invading robots, AUVs with different abilities form several hunting teams to hunt down the intruding robots. The dynamic encircling alliance constructed by the method proposed in the research can make full use of the sailing ability of every AUV, which will shorten the encircling time. The proposed algorithm is used in heterogeneous multi-AUV cooperative encircling and can select AUVs with similar capabilities to form an alliance so that each AUV in the round-up team travels at a similar speed. However, the distance competition mechanism ignores the sailing capabilities of AUVs, so the alliance may have more AUVs with widely varying capabilities. As with the barrel principle, the ability to encircle is limited by the AUV with less ability. The proposed time competition mechanism is more suitable for alliance generation of heterogeneous AUVs. The simulation result shows that the proposed time-based competition mechanism completes the encircling of invading robots at a lower distance cost.

(2) When encircling, the AUV should track the intruding robot as fast as possible. Combining the reliability function and the GBNN model can guide the AUV to immediately hunt down the intruding robot after forming the intruding robot's encircling alliance, and avoid the inefficient path. In the encircling problem, the algorithm combining the belief function and the GBNN model significantly outperforms the single GBNN model. The reliability function locally affects the neuron activity value, bringing a more definite forward goal for AUV. As the encircling step progresses, the neural activity value of the reliability function gradually decreases, and the GBNN model plans a safe obstacle avoidance path for the AUV to guarantee the success of the encircling.

Through comparative simulation analysis, the time-based encircling coalition generation algorithm proposed in this research is better than the distance-based competition mechanism, and the comprehensive path planning algorithm is also better than the single GBNN model.

3.4.2 Simulation Encircling in a 3-D Environment

This section presents a simulation study of a cooperative encircling intrusion robot with eight heterogeneous AUVs in a 3-D underwater environment.

3.4.2.1 Simulation Analysis of Encircling in a 3-D Environment

As shown in Figure 3.12, the simulation underwater environment is assumed to be a cuboid with a length and width of -60 to 60 and a height of -16 to 60. The underwater environment is shown on a 1:100 scale. At the bottom of the environment,

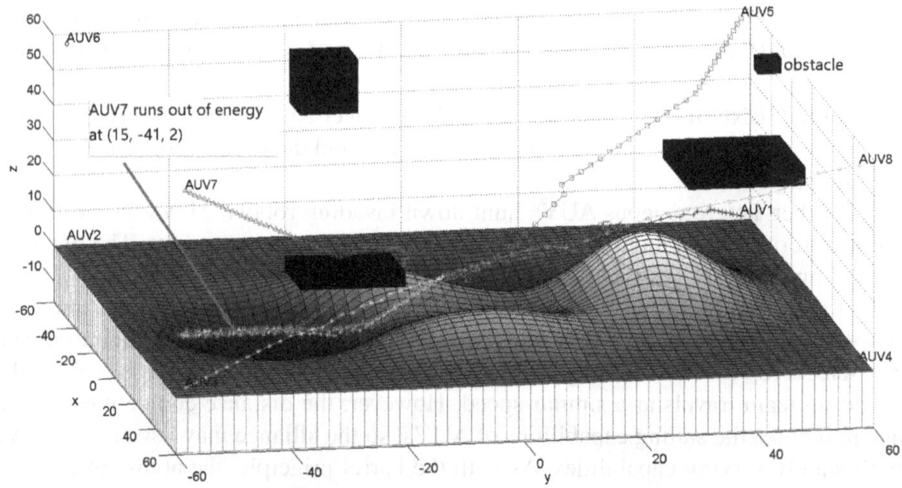

FIGURE 3.14 Simulation of the 3-D environment in distance-based competition mechanism (time step 78).

there are continuous hills on the seabed, the lowest height coordinate of the valley is about -15, and the height coordinate of the peak of the mountain is 20. Then, as shown in Figure 3.13, the position of the z-axis coordinate 0 is set to 0 meters above sea level. Taking this as a reference height, the height of the valley is -1,500 meters, and the height of the peak is 2,000 meters. The environment also sets up a number of obstacles represented by black blocks. There is an AUV in each of the 8 corners of the 3D environment. The invading robot is initially located at (20, 20, 30). The types of intruding robots and AUVs are the same as the simulation analysis in the 2-D case. In a 3-D environment, the AUV needs a long sailing distance to capture the intruding robot, so the safety distance also needs to be considered to prevent the loss of the AUV energy due to exhaustion. In the simulation, the parameters of the comprehensive reliability function and the GBNN path planning algorithm are set as:^{2} =0.01, v=200, ^{3} =3,R=1.8, c=0.1, m=10. Figure 3.12 and Figure 3.13 show the encircling process for this simulation analysis.

A time competition mechanism assigns encircling tasks to AUVs 1, 2, 3, 4, 5, and 8. AUV1 sailed 13,976 meters to the round-up occupancy point, and the pursuit distances of AUV2, AUV3, AUV4, AUV5, and AUV8 were 13,666, 10,024, 11,971, 14,092, and 12,634 meters, respectively. The AUV can avoid obstacles and hills on the ocean floor to hunt down intruding robots on a safe path. During the encircling, the AUV8 avoided collisions with obstacles. AUV1 and AUV4 avoid collisions with hills on the ocean floor. The invading robot tried its best to escape and chose the direction with the greatest escape probability, but the encircling AUV still surrounded it at the position of the grid (15, -30, 1), and the escape distance was 6491 meters.

3.4.2.2 Comparative Simulation Analysis of Distance-based Encircling Coalition Generation in a 3-D Environment

Since a single GBNN model cannot complete the tracking of the intruding robot in a 3-D environment, this section only conducts a comparative simulation analysis of different round-up alliance generation algorithms. For the same encircling simulation environment as section 4.2.1, encircling alliances are generated for AUVs based on the distance competition mechanism, and simulation analysis is carried out. The encircling alliance formed by AUVs is different from section 4.2.1. As shown in Figure 3.14, AUV7 replaced AUV2 in the encircling alliance. However, AUV7 is the slowest moving and has to chase hard to catch up with the invading robot. In addition, the safe distance of AUV7 is only 12,000 meters. If the AUV7 depletes energy during the encircling, the task mission will fail. Figure 3.14 shows the final stage of the encircling, where the AUV7 runs out of energy because the distance exceeds its safe distance. Therefore, since the original alliance generation algorithm only considers distance, it is not suitable to generate encircling alliances for heterogeneous AUVs in a 3-D environment.

Under a 3-D environment, when the abilities of each AUV in the encircling team are quite different, the efficiency is worse than that in the 2-D case. Due to the vastness of the environment, it is easier for invading robots to escape. Encircling alliance if

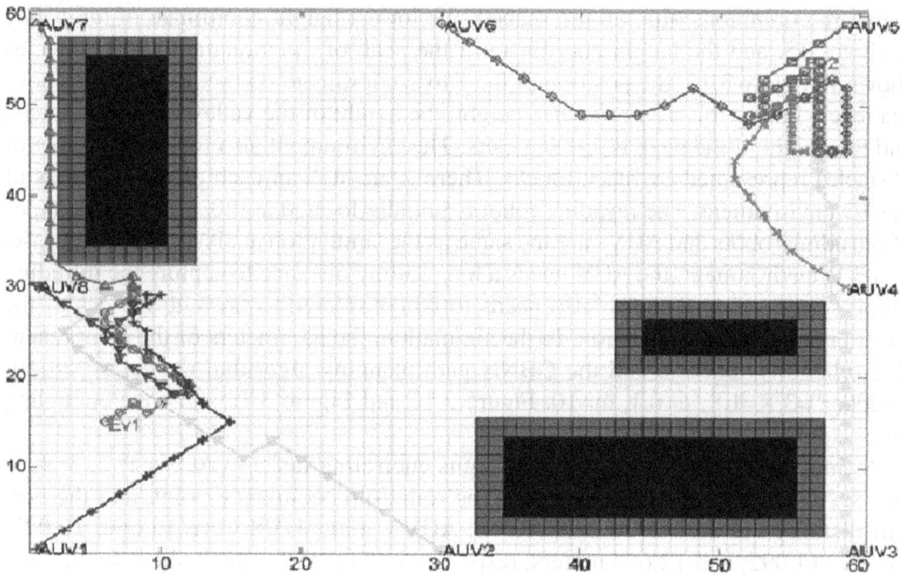

FIGURE 3.15 Encircling process by the improved GBNN model (complete encircling).

TABLE 3.4
Comparison of Encircling Simulation Results of Two Proposed Models

AUV number	Improved GBNN model	BNN model
1	3,814 m	4,040 m
2	4,506 m	4,506 m
3	4,347 m	5,046 m
4	3,573 m	3,926 m
5	2,831 m	3,097 m
6	4,619 m	4,434 m
7	3,455 m	4,120 m
8	3,028 m	3,434 m
AUVs' Average distance	**3,772 m**	**4,067 m**
Ev1	2,180 m	2,604 m
Ev2	2,307 m	2,573 m
Computation time	**2.9 s**	**311.2 s**

there is a slower AUV that cannot catch up with the invading robot, it will be difficult to form the final encircling. And if the safe distance of the AUV is not considered, the encircling may also fail. However, the proposed algorithm allocates the task to AUV teams with almost the same capabilities and focuses on the AUV's sailing safety distance and sailing speed, so that the AUV can complete the encircling faster and better, and improve the success rate.

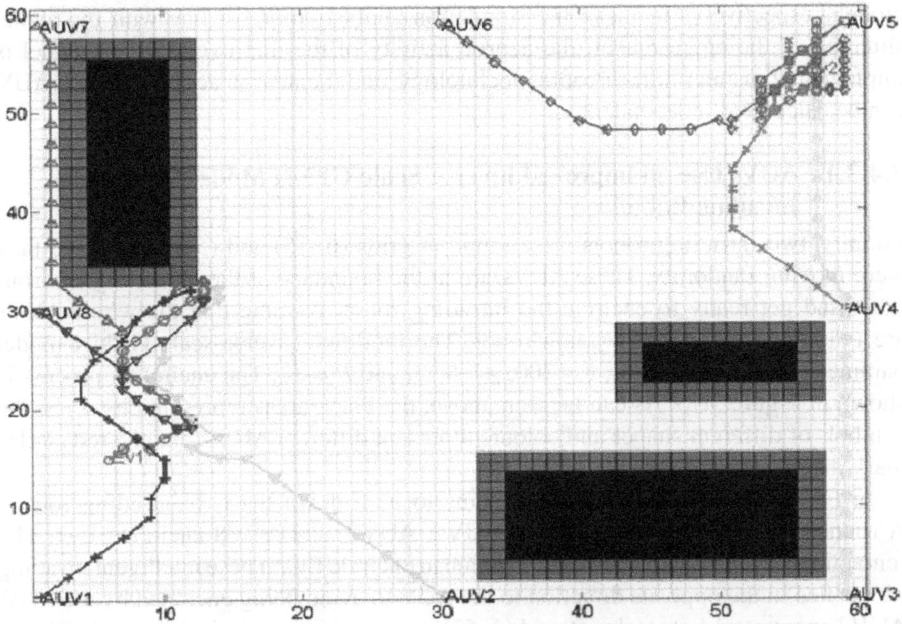

FIGURE 3.16 The encircling process using the BNN model (encircling completed).

3.4.3 Simulation of Pursuit Path Planning Algorithm Based on Improved GBNN Model

For the purpose of verifying the feasibility of the proposed improved GBNN model in AUV pursuit path control, several simulation analyses were performed using the *Matlab* platform. Then, simulation in a 2-D environment is carried out, and multiple random Monte Carlo simulations are carried out. The simulation research uses the improved GBNN and the BNN model to program the pursuit path for the AUV, and the results are compared and analyzed to show the advantages of the improved GBNN model. Although AUVs and intelligent intrusion robots are regarded as mass points, this section enlarges the obstacles by 2 grid cells to take into account their actual shapes and improve the safety of path planning. If the enlarged obstacle grid is the encroaching occupancy point of the intrusion robot, the AUV can also enter the occupancy point, while the intelligent intrusion robot is not allowed to enter the enlarged obstacle space.

In a 2-D environment, 8 AUVs encircle 2 intelligent intrusion robots. The size of the designed round-up environment is a two-dimensional grid environment of 60*60, and the simulation environment is displayed at a scale of 1:100. The perception range of the intelligent intrusion robot is a circular area with a diameter of 10 grids. When the intelligent intrusion robot detects that there is an encircling AUV around, it will adopt an intelligent escape strategy to avoid the encircling of the AUV. In the simulation study, the whole AUVs were twice as fast as the intelligent intrusion robot. The encircling task is achieved by an improved inverse-scale GBNN model and a BNN model, respectively. To make the simulation analysis more in line with the actual situation of the environment, the neuron activity of the improved GBNN model is implemented using multi-threading technology, and the neuron activity of each AUV can be calculated independently.

3.4.3.1 AUVs Use an Improved Inverse-Scale GBNN Model to Encircle Invading Robots

Initially, two intruding robots were found at grids (6, 15) and (56, 54), where they were moving randomly. These AUVs are at the boundary of the encircling environment and get ready to encircle the intruding robot. Assume that the whole AUVs are twice as fast as the invading robot. The modified inverse-scale GBNN model parameters are set to: $\beta = 1$, $v = 200$, $\gamma = 5.51$, and $R = \sqrt{2}$. The encircling process is shown in Figure 3.15. As can be seen above, the black areas represent obstacles, the symbols of different shapes and colours represent different AUVs, and the red circles mark the intruding robots.

In Figure 3.15, 8 AUVs form 2 teams to hunt down the two intruding robots. A round-up team containing AUV1, AUV2, AUV7, and AUV8 encircled Ev1. The remaining AUVs formed another AUV team to encircle Ev2. These intelligent invading robots flee in the opposite direction to the AUV to avoid being encircled by the AUV. AUV7 encountered obstacles at grid (2, 57), using the improved inverse-scale GBNN model, avoiding collisions with obstacles. In addition, the improved inverse-scale GBNN model makes the AUV gradually approach its encircling occupancy point

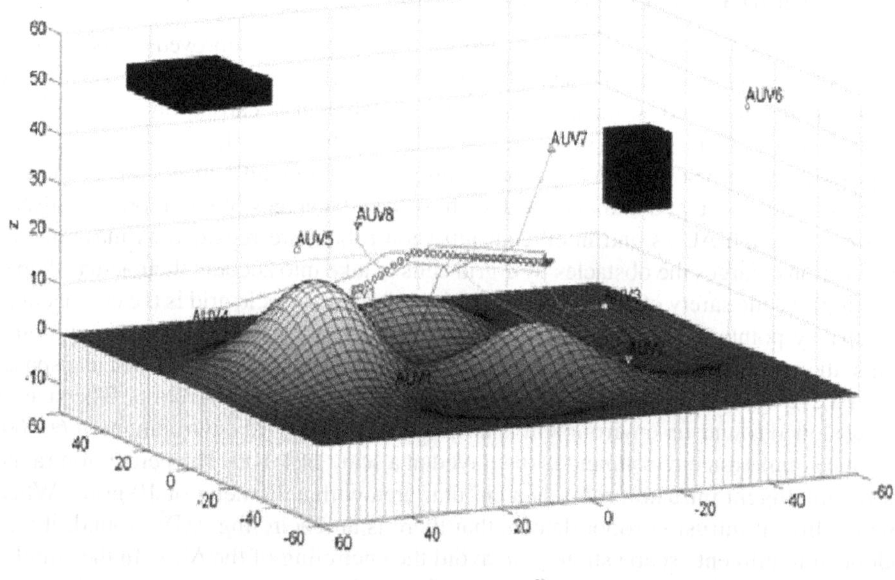

FIGURE 3.17 The encircling simulation process using the improved inverse-scale GBNN model (task completed).

to complete the round-up task. Ev1 and Ev2 are bounded on grids (8, 29) and (57, 55) respectively. A total of 40 encircling time steps were performed for this simulation. The second column of Table 3.4 lists the pursuit total distance cost of the AUV, and the escape distance of the invading robot.

3.4.3.2 AUV Uses the BNN Model to Encircle Invading Robots

For the same hunting environment setting above, based on the same hunting alliance generation algorithm, the AUV uses the BNN model to plan the hunting path and carry out corresponding simulation research. In the simulation, the parameters of the BNN model are set as: $B = D = 1$, $A = 25$, $u = 1$, and $R = 2$. Figure 3.16 shows the encircling results. The same coalition as in the simulation analysis of the improved GBNN model is constructed between AUVs. The encircling task is also accomplished using the BNN model, but with more hunting steps and more running time. Ev1 and Ev2 are enclosed in grids (13, 32) and (57, 52), respectively, and the simulation runs for 45 encircling time steps. Because the neuron activity update of the BNN model is slower than the improved GBNN model, AUV6 occupies the occupancy site for Ev2 later, and Ev2 escapes further, requiring more time. Ev1 also showed the same.

Through the improved GBNN model, the average sailing distance of the AUV is 3772 meters, completing the encircling task of two invading robots, while the average sailing distance of the BNN model is more, which is 4076 meters. According to related

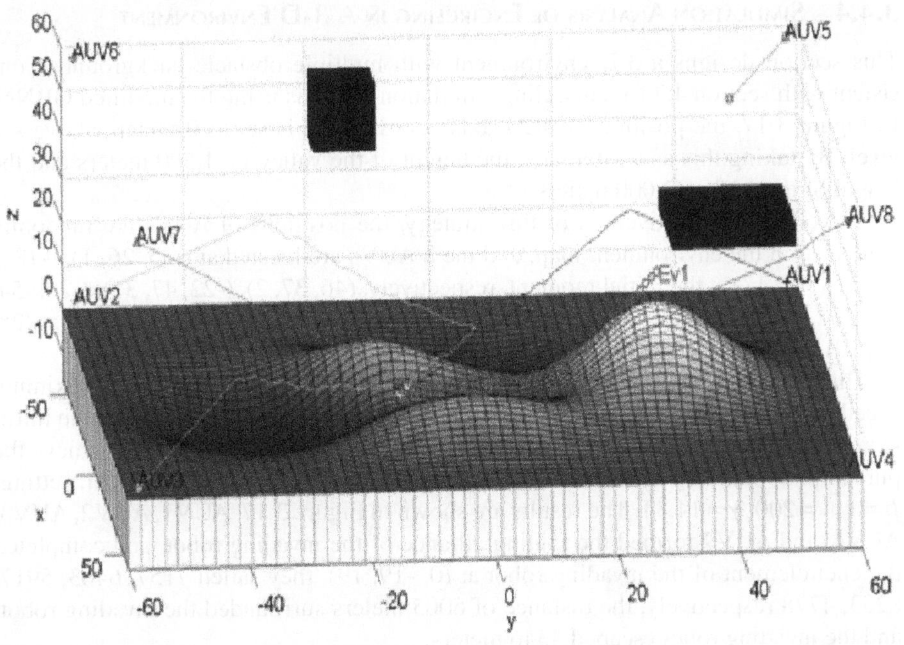

FIGURE 3.18 The encircling simulation process using the improved inverse-scale GBNN model (task completed).

research reports, the neuron activity of the BNN model changes rapidly and is suitable for real-time path planning required for quick pursuit. However, the improved GBNN model provides AUVs with a more direct and accurate pursuit direction, resulting in a shorter encircling distance. Technical improvements to the GBNN model to improve its dynamic performance to provide correct path information.

What's more, the improved GBNN model will not take much computation time, and the simulation program only needs 2.9 seconds to run. While the BNN model needs to solve the differential equation of every neuron, the computational complexity is high. For the same round-up simulation, the program needs to run for 311.2 seconds. In the simulation analysis, except for the different subroutines that calculate neuron activity, the other procedures are the same. Therefore, in terms of computational efficiency, the BNN model is very time-consuming and would be more difficult to apply in a 3-D underwater environment, because there are many neurons in a 3-D simulation environment to construct relevant information about the environment. With so many neurons requiring computation, the increased computational complexity of the activity of a single neuron will inevitably lead to a very time-consuming path planning process.

In summary, the improved GBNN model has good real-time path planning performance in a rapidly changing environment, improves the encircling efficiency, and is computationally efficient without sharing neuron activity information between AUVs.

3.4.4 Simulation Analysis of Encircling in a 3-D Environment

This section designs a 3-D environment with multiple obstacle backgrounds consistent with section 4.2 for encircling simulation analysis using the modified GBNN. In Figure 3.17, the position of the z-axis coordinate 0 is set to 0 meters above sea level. By taking this as a reference, the height of the valley is -1,500 meters and the height of the peak is 2,000 meters.

To verify the effectiveness of this strategy, the positions of AUVs are randomly generated in the environment map, and the 8 AUVs are located at (-25, 26, 1), (-18, -29, 3), (30, -47) at the initial moment, respectively. (46, 37, 2), (-22, 47, 31), (-20, -54, 51), (32, -36, 34), (12, 17, 27), while the invasion the robot is initially located at (28, 10, 10).

Through simulation, AUV3, AUV5, AUV6, and AUV7 have the same maximum speed as the intrusion robot, while the remaining AUVs are twice as fast as the intrusion robot. The simulation uses the inverse-proportional GBNN model to achieve the pursuit path planning of the AUVs. For the parameter settings of the model, setting: $\beta = 1$, $\nu = 200$, $\gamma = 14.34$. The results are shown in Figure 3.17. AUV1, AUV2, AUV3, AUV4, and AUV7 formed the raiding alliance of the invading robot, and completed the encirclement of the invading robot at (0, -19, 19), they sailed 7157, 6403, 5917, 8253, 4778 respectively, the distance of 6005 meters surrounded the invading robot, and the invading robot escaped 4346 meters.

It is easy to see that the proposed encircling strategy can also finish the task in the case of random AUV and invading robot positions.

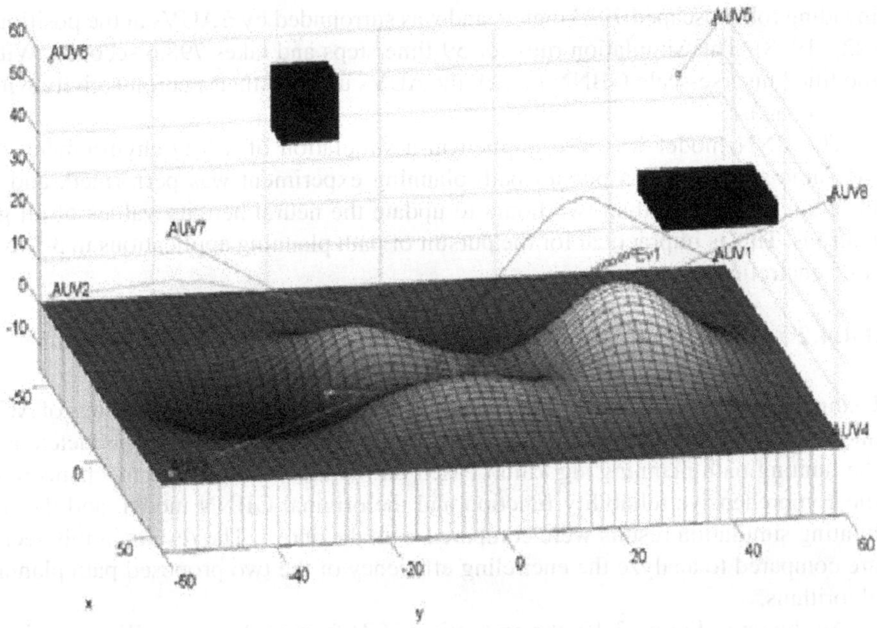

FIGURE 3.19 The encircling simulation process using the synthetic reliability function and the original GBNN model (task completed).

3.4.4.1 Simulation Analysis of Pursuit of Improved GBNN Algorithm

In the case of the same 3-D underwater background environment setting as before, a comparative simulation of the pursuit path planning is carried out. The encircling AUVs are located in 8 corners of the simulation environment. The invading bot is initially located at grid (20, 30, 34). Under these experiments, AUV3, AUV5, AUV6, and AUV7 have the identical maximum speed as the invading robot, while other AUVs are twice as fast as the invading robot. The simulation program uses the inverse-proportional GBNN model to complete the pursuit path planning, and the parameters of the model are set as same as in the previous section.

Figure 3.18 shows the process of AUV encircling the intruding robot in this simulation analysis. Encircling tasks are firstly allocated to AUV1, AUV2, AUV4, AUV5, AUV7, and AUV8. Because the invading robot gradually approached AUV3 and moved away from AUV5 when escaping, which made the distance between the intruding robot and AUV5 exceed the distance between it and AUV3, AUV5 was substituted for AUV3 at the 10th time step, which was replaced by AUV3 continue to pursuit invading robots. AUV5 sailed 1732 meters at this point and stopped at grid (-48, 48, 48). During the encircling, AUVs can prevent collisions with hills and obstacles, and once the invading robot adopts an evasion strategy, adjusts the escape direction, and runs to the grid in the other direction, the AUVs quickly change their pursuit direction. Finally, the distances tracked by AUV1, AUV2, AUV3, AUV4, AUV7, and AUV8 are 12678, 12932, 7368, 9274, 7781, and 10445 meters, respectively. The invading robot escaped 7923 meters and was surrounded by 6 AUVs at the position of (22, -19, 8). This simulation runs for 59-time steps and takes 798.6 seconds. With a modified inverse-scale GBNN model, the AUVs updated their neuron activity within 13.5 seconds.

The BNN model was also applied in a simulation of a 3-D environment with the same setting, and a pursuit path planning experiment was performed, and the program costs more than two hours to update the neural activity values of all grid neurons. This is impractical for the pursuit of path planning applications in 3-D real-time encircling.

3.4.4.2 Pursuit Simulation Analysis of Pursuit Path Planning Based on Synthetic Belief Function and GBNN Model

Using the same hunting environment and the location and capability settings of AUVs and intruding robots, the encircling process simulation analysis was conducted using the hunting path planning algorithm based on the time competition mechanism and the comprehensive reliability function and the original GBNN model, and the final hunting simulation results were compared with the above. The results in this section are compared to analyze the encircling efficiency of the two proposed path planning algorithms.

As shown in Figure 3.19, the encircling AUV formed the same alliance as in the previous section, and finally completed the task. AUV5 is also replaced by AUV3 at the 10th time step. However, the comprehensive reliability function and the pursuit path planning algorithm of the original GBNN model are different from the improved inverse-proportional GBNN model, and the pursuit direction is slightly worse, which makes the pursuit distance longer, and causes the intrusion robot to be the slightly different direction of escape. The distances tracked by AUV1, AUV2, AUV3, AUV4, AUV7, and AUV8 are 13193, 13248, 6413, 11291, 8962, and 13590 meters, respectively. The invading robot escaped 9055 meters and was surrounded by 6 AUVs at the position (15, -23, 3). Therefore, the efficiency of the comprehensive path planning algorithm is lower than that of the improved inverse-proportional GBNN model, which makes the intruding robot escape a longer distance, but it can also complete the pursuit path planning of multiple AUVs with higher encircling efficiency. Several improvements of GBNN have greatly improved its dynamic characteristics, making it more suitable for dynamic tracking environments similar to encircling.

3.5 CONCLUSION

This chapter summarizes the task assignment algorithms and applies a time-based improved Self Organized Mapping algorithm to encircling task assignments. The path planning for pursuing the intrusion devices is also a key factor to ensure fast encirclement. After the review of path planning algorithms, an algorithm based on an improved GBNN model is proposed to achieve fast and dynamic tracking. The simulation study is performed both in two-dimensional. Then, these results show that the proposed algorithms can efficiently fulfill the encircling task. Through the comparison simulations, its efficiency is better than the distance-based task assignment and the BNN algorithms. Firstly, the escape intelligence of invading robots is discussed. Then a time competition mechanism is adopted to form an efficient dynamic encircling alliance for AUVs. In the pursuit process, an algorithm combining GBNN and reliability function algorithm and an improved inverse-proportional GBNN path planning algorithm was designed for AUV, providing efficient pursuit path planning for AUV. The simulation study verifies the feasibility and efficiency of the encircling coalition generation algorithm and real-time path planning method.

In addition, some comparative simulation analysis is carried out. At first, the encircling coalition generation algorithm is compared with the original distance competition algorithm. The results show that the alliance generation method based on the time competition mechanism makes the AUV capabilities of the encircling teams the same, and improves efficiency. Then, the simulation study compares the algorithm of integrating GBNN and reliability function with a single GBNN model. The integrated algorithm reduces the lag time of the AUV's pursuit action and accelerates the AUV's tracking speed. Finally, the improved GBNN algorithm and BNN algorithm are respectively used in the cooperative encircling task of homogeneous AUVs. A simulation study shows that the improved GBNN algorithm has better dynamic performance than the BNN model, the planned pursuit path is more accurate, and there is no delay in the propagation of neural activity, which makes the pursuit fast and accurate enough, and reduces the pursuit path. More importantly, the GBNN algorithm has low computational complexity and is faster than the BNN algorithm. It does not need to share neuron activity between AUVs and is more suitable for cooperative encircling in an underwater weak communication environment. Moreover, the final 3-D simulation comparison shows that the improved GBNN algorithm also has a higher encircling efficiency than the algorithm combining the original GBNN and the belief function.

ACKNOWLEDGMENT

This work was supported in part by the National Natural Science Foundation of China (52001195). The authors would also like to thank the editors for their hard work.

REFERENCES

[1] D. R. Blidberg, "The Development of Autonomous Underwater Vehicles (AUV); A Brief Summary," *IEEE ICRA,* 2001.

[2] W. Wang, J. Qi, H. Zhang, and G. Zong, "A Rapid Hunting Algorithm for Multi Mobile Robots System," in *Industrial Electronics and Applications, 2007. ICIEA 2007. 2nd IEEE Conference on*, 2007.

[3] Y. Wang, L. Dong, and C. Sun, "Cooperative Control for Multi-Player Pursuit-Evasion Games With Reinforcement Learning," *Neurocomputing,* vol. 412, pp. 101–114 ,2020.

[4] Y. Song, Y. Li, C. Li, and X. Ma, "Mathematical Modeling and Analysis of Multirobot Cooperative Hunting Behaviors," *Journal of Robotics,* vol. 2015, pp. 11–19, 2015.

[5] J. Ni and S. X. Yang, "Bioinspired Neural Network for Real-Time Cooperative Hunting by Multirobots in Unknown Environments," *IEEE Transactions on Neural Networks,* vol. 22, no. 12, pp. 2062–2077, 2011.

[6] D. P. Williams, "On Optimal AUV Track-spacing for Underwater Mine Detection," in *IEEE International Conference on Robotics & Automation*, 2010.

[7] D. Zhu and Z. Huang, "A Cooperative Hunting Algorithm of Multi-AUV in 3-D Dynamic Environment," in *Control & Decision Conference*, 2015.

[8] C. Xiang, Z. Huang, and D. Zhu, "AUV Cooperative Hunting Algorithm Based on Bio-inspired Neural Network for Path Conflict State," in *2015 IEEE International Conference on Information and Automation*, 2015.

[9] M. Chen and D. Zhu, "A Novel Cooperative Hunting Algorithm for Inhomogeneous Multiple Autonomous Underwater Vehicles," *IEEE Access,* vol. 6, pp. 7818–7828, 2018.

[10] J. Ni, L. Yang, L. Wu, and X. Fan, "An Improved Spinal Neural System-Based Approach for Heterogeneous AUVs Cooperative Hunting," *International Journal of Fuzzy Systems,* 2017.

[11] Z. Yan, C. Zhang, W. Tian, and M. Zhang, "Formation Trajectory Tracking Control of Discrete-time Multi-AUV in a Weak Communication Environment," *Ocean Engineering,* vol. 245, pp. 110495–110508 2022, doi: 10.1016/J.Oceaneng.2021.110495

[12] X. Chen, P. Zhang, G. Du, and F. Li, "A Distributed Method for Dynamic Multi-robot Task Allocation Problems with Critical Time Constraints," *Robotics and Autonomous Systems,* vol. 118, pp. 31–46, 2019, doi: 10.1016/j.robot.2019.04.012

[13] A. Sahoo, S. K. Dwivedy, and P. S. Robi, "Advancements in the Field of Autonomous Underwater Vehicle," *Ocean Engineering,* vol. 181, pp. 145–160, 2019, doi: 10.1016/j.oceaneng.2019.04.011

[14] J. Hu and Q. B. Zhu, "A Multi-Robot Hunting Algorithm Based on Dynamic Prediction for Trajectory of the Moving Target and Hunting Points," *Acta Electronica Sinica,* vol. 39, no. 11, pp. 2480–2485, 2011.

[15] J. Li, Q. S. Pan, B. R. Hong, and M. H. Li, "Multi-robot Cooperative Pursuit Based on Association Rule Data Mining," in *International Conference on Fuzzy Systems & Knowledge Discovery*, 2009.

[16] W. Ming, F. Huang, W. Long, and J. Sun, "A Distributed Multi-Robot Cooperative Hunting Algorithm Based on Limit-cycle," *Microcomputer Information,* 2010.

[17] A. Khamis, A. Hussein, and A. Elmogy, "Multi-robot Task Allocation: A Review of the State-of-the-Art," in *Cooperative Robots and Sensor Networks 2015*, (Studies in Computational Intelligence, 2015, ch. Chapter 2, pp. 31–51.

[18] S. Keshmiri and S. Payandeh, "Multi-robot, Dynamic Task Allocation: A Case Study," *Intelligent Service Robotics,* vol. 6, no. 3, pp. 137–154, 2013, doi: 10.1007/s11370-013-0130-x

[19] G. A. Korsah, A. Stentz, and M. B. Dias, "A Comprehensive Taxonomy for Multi-robot Task Allocation," *International Journal of Robotics Research,* vol. 32, no. 12, pp. 1495–1512, 2013.

[20] R. Zlot and A. Stentz, "Market-based Multirobot Coordination for Complex Tasks," *International Journal of Robotics Research,* vol. 25, no. 1, pp. 73–101, 2006.

[21] M. Chen and D. Zhu, "Optimal Time-Consuming Path Planning for Autonomous Underwater Vehicles Based on a Dynamic Neural Network Model in Ocean Current Environments," *IEEE Transactions on Vehicular Technology,* vol. 69, no. 12, pp. 14401–14412, 2020, doi: 10.1109/tvt.2020.3034628

[22] Z. Zhu, B. Tang, and J. Yuan, "Multirobot Task Allocation Based on an Improved Particle Swarm Optimization Approach," *International Journal of Advanced Robotic Systems,* vol. 14, no. 3, pp. 1729–1739, 2017.

[23] Y. Chen, D. Yang, and J. Yu, "Multi-UAV Task Assignment With Parameter and Time-Sensitive Uncertainties Using Modified Two-Part Wolf Pack Search Algorithm," *IEEE Transactions on Aerospace and Electronic Systems,* 2018.

[24] B. Horling and V. Lesser, "A Survey of Multi-Agent Organizational Paradigms," *Knowledge Engineering Review,* 2005.

[25] C. C. Corona and D. A. Pelta, *Self-adaptation in Dynamic Environments – A Survey and Open Issues*, 2016.

[26] S. Sariel, T. Balch, and J. Stack, "Distributed Multi-AUV Coordination in Naval Mine Countermeasure Missions," *Robotics & Automation Magazine IEEE,* vol. 2, pp. 11–17, 2006.

[27] V. Singhal and D. Dahiya, "Distributed Task Allocation in Dynamic Multi-agent System," in *International Conference on Computing*, 2015.

[28] A. M. Elmogy, A. M. Khamis, and F. Karray, "Market-Based Framework for Mobile Surveillance Systems" *Springer Berlin Heidelberg,* vol. 7326, pp. 69–78, 2012.

[29] J. Li and R. B. Zhang, "Multi-Auv Distributed Task Allocation Based on the Differential Evolution Quantum Bee Colony Optimization Algorithm," *Polish Maritime Research,* vol. 24, no. s3, pp. 65–71, 2017, doi: 10.1515/pomr-2017-0106

[30] J. Lima, P. Costa, P. Costa, L. Eckert, and A. Nakano, "A* Search Algorithm Optimization Path Planning in Mobile Robots Scenarios," in *Central European Symposium on Thermophysics 2019 (CEST)*, 2019.

[31] Jian *et al.*, "Using Auction-based Task Allocation Scheme for Simulation Optimization of Search and Rescue in Disaster Relief," *Simulation Modelling Practice and Theory: International Journal of the Federation of European Simulation Societies,* vol. 82, pp. 132–146, 2018.

[32] V. Krishna and J. S. Bentley, "Auction Theory," *Vijay Krishna.*

[33] F. Zhang, "A Novel Hybrid Auction Algorithm for Multi-UAVs Dynamic Task Assignment," *IEEE Access,* vol. 8, no. 2169–3536, pp. 86207–88622, 2019.

[34] P. Caloud, W. Choi, J. C. Latombe, C. L. Pape, and M. Yim, "Indoor Automation with Many Mobile Robots," in *Intelligent Robots and Systems '90. 'Towards a New Frontier of Applications', Proceedings. IROS '90. IEEE International Workshop on*, 1990.

[35] B. P. Gerkey and M. J. Mataric, "Sold!: Auction Methods for Multirobot Coordination," *IEEE Transactions on Robotics and Automation,* vol. 18, no. 5, pp. 758–768, 2002, doi: 10.1109/tra.2002.803462

[36] J. Jiang, J. H. Yan, and X. Z. Zang, "Multi-robot Systems Task Deadlock Resolution Based on Combinatorial Auction," *Computer Engineering & Applications,* vol. 44, no. 4, pp. 1–3, 2008.

[37] W. N. Huang, J. W. Gong, and P. H. Wang, "Multi-robot Task Allocation Method Based on GA," *Computer Simulation,* vol. 23, no. 11, pp. 164–167, 2006.

[38] D. H. Lee, S. A. Zaheer, and J. H. Kim, "A Resource-Oriented, Decentralized Auction Algorithm for Multirobot Task Allocation," *IEEE Transactions on Automation Science & Engineering,* vol. 12, no. 4, pp. 1469–1481, 2015.

[39] C. Qiao, Y. Dong, Y. Jian, and L. Shen, "An Auction-Based Multiple Constraints Task Allocation Algorithm for Multi-UAV System," in *2016 International Conference on Cybernetics, Robotics and Control (CRC)*, 2016.

[40] Z. Zhang, W. Jie, X. Dong, and Y. Meng, "Task Allocation of Multi-AUVs Based on Innovative Auction Algorithm," in *2017 10th International Symposium on Computational Intelligence and Design (ISCID)*, 2017.

[41] J. Shi, Z. Yang, and J. Zhu, "An Auction-based Rescue Task Allocation Approach for Heterogeneous Multi-robot System," *Multimedia Tools and Applications,* vol. 79, pp. 14529–14538, 2018.

[42] M. U. Arif and S. Haider, "An Evolutionary Traveling Salesman Approach for Multi-Robot Task Allocation," in *9th International Conference on Agents and Artificial Intelligence*, 2017.

[43] T. Wang, F. U. Xiao-Wei, and X. G. Gao, "Cooperative Task Assignment for Heterogeneous Multi-UAVs Based on Improved Genetic Algorithm," *Fire Control & Command Control,* vol. 8, pp. 11–19, 2013.

[44] J.-J. Li, R.-B. Zhang, and Y. Yang, "Multi AUV Intelligent Autonomous Learning Mechanism Based on QPSO Algorithm," in *Intelligent Computing Theories and Methodologies*, (Lecture Notes in Computer Science, 2015, ch. Chapter 6, pp. 60–67.

[45] S. M. Zadeh, D. Powers, K. Sammut, and A. M. Yazdani, "A Novel Versatile Architecture for Autonomous Underwater Vehicle's Motion Planning and Task Assignment," *Soft Computing,* vol. 22, no. 5, pp. 1687–1710, 2016.

[46] J. P. Wang, Y. Gu, and X. M. Li, "Multi-robot Task Allocation Based on Ant Colony Algorithm," *Journal of Computers,* vol. 7, no. 9, pp. 2160–2167, 2012.

[47] Z. Haifeng, "An Improved Ant Colony Algorithm Combined with Genetic Algorithm and Its Application in Image Segmentation," *Journal of Cosmetic & Laser Therapy Official Publication of the European Society for Laser Dermatology,* vol. 15, no. 1, pp. 25–33, 2013.

[48] R. Eberhart and J. Kennedy, "A New Optimizer Using Particle Swarm Theory," in *Mhs95 Sixth International Symposium on Micro Machine & Human Science*, 2002.

[49] LU, P. Qian, D. Cheng-Sheng, H. Yuan-Ming, and Rui-xiang, "Task Allocation of Multi-AUV Based on Multi-objective Discrete Particle Swarm Optimization and Simulated Annealing," in *The 32nd China Control and Decision-making Conference*, 2020.

[50] J. Li, R. Zhang, and Y. Yu, "Multi-AUV Autonomous Task Planning Based on the Scroll Time Domain Quantum Bee Colony Optimization Algorithm in Uncertain Environment," *Plos One,* vol. 12, no. 11, pp. e0188291, 2017.

[51] S. Mahmoudzadeh, D. M. W. Powers, K. Sammut, A. M. Yazdani, and A. Atyabi, "Hybrid Motion Planning Task Allocation Model for AUV's Safe Maneuvering in a Realistic Ocean Environment," *Journal of Intelligent & Robotic Systems,* vol. 7, no. 2, pp. 265–282, 2018.

[52] S. Saravanan, K. C. Ramanathan, M. M. Ramya, and M. N. Janardhanan, "Review on State-of-the-art Dynamic Task Allocation Strategies for Multiple-robot Systems," *Industrial Robot,* vol. 47, no. 6, pp. 929–942, 2020.

[53] X. Chen, Z. Ping, L. Fang, and G. Du, "A Cluster First Strategy for Distributed Multi-robot Task Allocation Problem with Time Constraints *," in *2018 WRC Symposium on Advanced Robotics and Automation (WRC SARA)*, 2018.

[54] M. Elango, S. Nachiappan, and M. K. Tiwari, "Balancing Task Allocation in Multi-robot Systems Using K-means Clustering and Auction Based Mechanisms," *Expert Systems with Applications,* vol. 38, no. 6, pp. 6486–6491, 2011.

[55] P. Ghassemi and S. Chowdhury, "Decentralized Task Allocation in Multi-Robot Systems via Bipartite Graph Matching Augmented with Fuzzy Clustering," *The ASME 2018 International Design Engineering Technical Conferences & Computers and Information in Engineering Conference, 2018.*

[56] H. Huang, D. Zhu, and F. Ding, "Dynamic Task Assignment and Path Planning for Multi-AUV System in Variable Ocean Current Environment," *Journal of Intelligent and Robotic Systems,* vol. 74, pp. 999–1012, 2014.

[57] A. Zhu and S. X. Yang, "An Improved SOM-based Approach to Dynamic Task Assignment of Multi-robots," in *Intelligent Control and Automation (WCICA), 2010 8th World Congress on*, 2010.

[58] D. Zhu, X. Cao, B. Sun, and C. Luo, "Biologically Inspired Self-Organizing Map Applied to Task Assignment and Path Planning of an AUV System," *IEEE Transactions on Cognitive and Developmental Systems,* vol. 10, no. 2, pp. 304–313, 2018, doi: 10.1109/tcds.2017.2727678

[59] D. Zhu, H. Huang, and S. X. Yang, "Dynamic Task Assignment and Path Planning of Multi-AUV System Based on an Improved Self-Organizing Map and Velocity Synthesis Method in Three-Dimensional Underwater Workspace," *IEEE Transactions on Cybernetics,* vol. 43, no. 2, pp. 504–14, Apr 2013, doi: 10.1109/TSMCB.2012.2210212

[60] A. Zhu and S. X. Yang, "A Neural Network Approach to Dynamic Task Assignment of Multirobots," *Transactions on Neural Networks,* vol. 17, no. 5, pp. 1278–1287, 2006, doi: 10.1109/tnn.2006.875994

[61] D. Zhu, Y. Liu, and B. Sun, "Task Assignment and Path Planning of a Multi-AUV System Based on a Glasius Bio-Inspired Self-Organising Map Algorithm," *The Journal of Navigation,* vol. 71, no. 2, pp. 482–496, 2018.

[62] X. Cao and A. L. Yu, "Multi-AUV Cooperative Target Search Algorithm in 3-D Underwater Workspace," *Journal of Navigation,* vol. 70, no. 6, pp. 1–19, 2017.

[63] Y. Carreno, E. Pairet, Y. Petillot, and R. P. A. Petrick, "A Decentralised Strategy for Heterogeneous AUV Missions Via Goal Distribution and Temporal Planning," in *30th International Conference on Automated Planning and Scheduling, ICAPS 2020, October 26, 2020–October 30, 2020,* Nancy, France, 2020, vol. 30: AAAI press, in Proceedings International Conference on Automated Planning and Scheduling, ICAPS, pp. 431–439.

[64] J. Wang, G. Jia, H. Xin, and Z. Hon, "Research on Dynamic Task Allocation Method of Heterogeneous Multi-UAV Based on Consensus Based Bundle Algorithm," in *2020 Chinese Automation Congress (CAC)*, 2020.

[65] L. Zhang, L. Zhang, and S. Liu, "Role-based Collaborative Task Planning of Heterogeneous Multi-autonomous Underwater Vehicles," *International Journal of Advanced Robotic Systems,* vol. 16, no. 3, pp. 1–7, 2019.

[66] T. Kohonen, "The Self-organizing Map," *IEEE Proc Icnn,* vol. 1, no. 1–3, pp. 1–6, 1990.

[67] V. V. Maneev and M. V. Syryamkin, "Optimizing the A* Search Algorithm for Mobile Robotic Devices," *IOP Conference Series Materials Science and Engineering,* vol. 516, no. 1, p. 012054, 2019.

[68] C. H. Chang, J. T. Zou, and K. L. Su, "Target Searching of Mobile Robots Using Improved A* Searching Algorithm," *Applied Mechanics & Materials,* vol. 300–301, pp. 519–522, 2013.

[69] A. Bakdi, H. Abdelfetah, H. Boutamai, A. Maoudj, and B. Bouzouia, "Optimal Path Planning and Execution for Mobile Robots Using Genetic Algorithm and Adaptive Fuzzy-logic Control," *Robotics and Autonomous Systems,* vol. 89, no. 1, pp. 95–109, 2017.

[70] K. Jiang and C. G. Li, "Path Planning Based on Fuzzy Logic Algorithm for Robots in Hierarchical Control," *Applied Mechanics & Materials,* vol. 644–650, pp. 701–704, 2014.

[71] J. Wang, "A Simulated Annealing Algorithm and Grid Map-Based UAV Coverage Path Planning Method for 3D Reconstruction," *Electronics,* vol. 10, pp. 853–869, 2021.

[72] S. Capraz, H. Azyikmis, and A. Ozsoy, "An Optimized GPU-Accelerated Route Planning of Multi-UAV Systems Using Simulated Annealing," *EJournal Publishing,* no. 3, 2020.

[73] L. P. Behnck, D. Doering, C. E. Pereira, and A. Rettberg, "A Modified Simulated Annealing Algorithm for SUAVs Path Planning," *IFAC PapersOnLine,* vol. 48, no. 10, pp. 63–68, 2015.

[74] W. Fang, C. Wang, and R. Zhao, "Application of Simulated Annealing Particle Swarm Optimization in Complex Three-dimensional Path Planning," *Journal of Physics: Conference Series,* vol. 1873, no. 1, pp. 012077–012089, 2021.

[75] X. U. He-Qun, "PID Parameters Optimization Based on Improved Ant Colony Algorithm," *Manufacturing Automation,* vol. 4, pp. 1–4, 2011.

[76] S. A. Mokhtari, "FOPID Control of Quadrotor Based on Neural Networks Optimization and Path Planning Through Machine Learning and PSO Algorithm," *International Journal of Aeronautical and Space Sciences,* vol. 23, no. 3, pp. 567–582, 2022.

[77] I. Sung, B. Choi, and P. Nielsen, "On the Training of a Neural Network for Online Path Planning with Offline Path Planning Algorithms," *International Journal of Information Management,* vol. 57, pp. 102142–102119, 2020.

[78] J. Immanuel, A. Suryadibrata, and J. C. Young, "Robotic Path Planning Using Evolutionary Neural Network," in *International Workshop on Advanced Imaging Technologies 2020 (IWAIT 2020)*, 2020.

4 Safety Management Based on Machine Learning for Identifying Unexpected Faults in Autonomous Underwater Vehicles

Edwin Ramirez-Asis, William Dextre-Martinez, and D. Saravanan

4.1 INTRODUCTION

Ocean platforms for ocean surveys and underwater structural management are becoming more popular due to geopolitical and energy-related factors. Scuba diving or laboring for more than an hour may be challenging in coastal areas prone to construction and human activity. Because of this, several governments across the globe are working to build autonomous ships and underwater robots that can take the place of human workers (Cho et al., 2020). Water management is the process of planning, generating, distributing, and managing water resources efficiently. Many elements of human life are affected by these behaviors, including food production and consumption. The procurement of smart water solutions is aided by smart water management. AUVs, or autonomous underwater vehicles, may be utilized to perform mine surveys without the requirement for a human operator on the shore or the seabed. To perform its mission, the autonomous underwater vehicle (AUV) follows a predefined path (Nie et al., 2020). AUV sensors, such as an optic camera, a side scan sonar, and an ultrasonic camera, can be used to detect and locate underwater objects. Figure 4.1 depicts the Satellite and Vehicle communication.

Even in metropolitan settings, where it is difficult to keep track of water usage, water conservation is an important part of water management. Autonomous underwater vehicles (AUVs) have gained prominence in the oil and gas sector and other subsea applications in recent years. With the fast advancement of ocean engineering, the control of AUVs has garnered a lot of attention (Yan and Yu, 2018). In the field

DOI: 10.1201/9781003320074-4

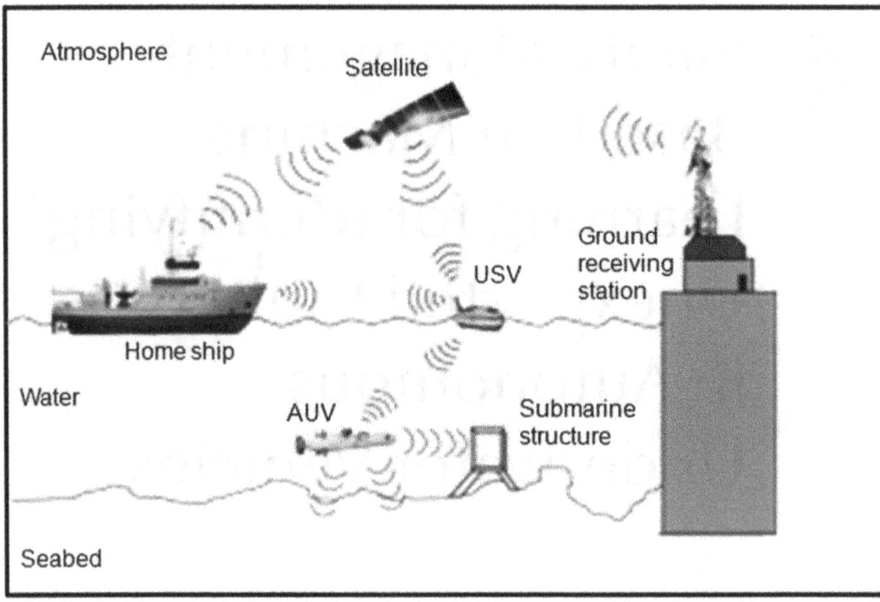

FIGURE 4.1 Satellite and vehicle communication.

of maritime engineering and control, AUV trajectory tracking control is a crucial issue to solve. Subsea circumstances such as model non-linearities, unknown hydrodynamic coefficients, and disturbances limit the capabilities of AUVs. Hence, the development of effective AUV controllers is a top priority for the maritime sector (Wang et al., 2020). Ocean energy (such as heat, salinity, wave, current, and tidal energy) other renewable energy sources (such as wind and solar energy that ocean's surface reflects) have enormous energy storage. Renewable energy may offer a steady source of energy to UUVs, allowing them to extend their range and perform high-frequency sampling. While renewable energy is less polluting, it also has a greater potential to protect the ocean ecosystem from contamination (Kong et al., 2020). Because of this, most people agree that the main source of electricity for the next generation of UUVs will be renewable energy. Wind, solar, wave, thermal, and other forms of renewable energy are now the most often used sources of power for UUVs. The features of each of these energy sources are distinct. Wind energy is a typical renewable energy source because of its high energy density and ease of access (up to 50 W/m^2 on average) (Wang et al., 2021). "Ocean Aero Submarine" is the name given to a conventional wind-powered UUV. But because ocean wind only blows at the ocean's surface, the UUVs have to wait a long time to refuel. Solar energy is more cost-effective than wind power has a far greater energy density (up to 168 W/m^2). The output power of SAUV-II, for example, may range from 6 to 18 watts (Vu et al., 2021). As a result, when compared to other renewable energy sources, solar energy has a major advantage. Solar UUVs, on the other hand, have to wait longer on the sea surface for charging since they are often charged during the day (Chen et al., 2018).

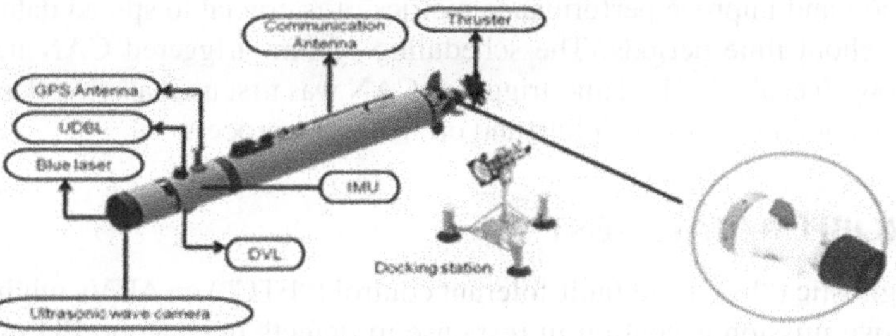

FIGURE 4.2 Architecture of AUV.

These UUVs are more vulnerable to biological contamination, making them unsuitable for covert operations. Wave energy, the third most popular ocean energy source, generally has a density in energy of 2 kW/m of wave height. With that said, shallow areas are more suited to the wave's exponential decay (Londhe and Patre, 2019). An example of wave energy UUV is the Wave Glider SV2. Unmanned undersea vehicles, known as autonomous underwater vehicles (AUVs), may be used to conduct mine surveys along the shore or on the seabed without the need for a human operator. The AUV is self-contained and follows predetermined routes to complete its task. It is possible to identify and locate underwater objects with the use of AUV sensors, such as an optic camera, a side scan sonar, and an ultrasonic camera. Fig. 4.2 depicts the Architecture of AUV (Rangel et al., 2020).

Academic and industrial researchers have concentrated on AUV control systems' use of a shared communication network. One of the concerns is the network-induced delay that happens during data transfer. The limited bandwidth of a shared communication network must meet each function node's communication demands (Liu et al., 2020). A result of fierce competition for limited network capacity, data packet failures and network delays may occur in the AUV control network. The AUV's performance may be impacted if the control system cannot gather the required data quickly enough to calculate control actions and communicate them to rudder motors (Xia et al., 2021). The network-induced delay occurs when several data transmission requests compete for limited network capacity. The network-induced delay occurs because only one of these information transmission needs may contact the network at a time (Antonelli, 2018). Network scheduling is used by the AUV control system to distribute limited network capacity among all information transmission demands, reducing or eliminating network congestion. Events begin CAN communication. Each bit in the standard CAN have a priority and competes for bandwidth using non-destructive bit-wise arbitration media access control (Ji et al., 2021). While high priority data may be sent quickly, this medium access control technique can cause low priority data on crowded networks to experience large transmission delays and erratic transmission behavior. The allocated data is sent through the CAN protocol at predetermined intervals (Zhao et al., 2021). The first stage in time-triggered CAN implementation is to create network information transfer schedules. Network scheduling commonly uses data transmission matrices. To satisfy the demand for data transmission and improve performance metrics, it is crucial to spread data transmission throughout time periods. The scheduling of time-triggered CAN networks is listed below (Ji et al., 2021). Time-triggered CAN was first used a network scheduling technology combining a scheduling and optimization process.

4.2 PROBLEM STATEMENT

Fault diagnostic ("FD") and fault-tolerant control ("FTC") on AUVs might increase and improve mission execution in response to defects in concept. FD in real-time should be a proactive measure to uncover system flaws and provide FTC early warning of the magnitude and location of any problems. Many years of AUV FD research have been dedicated to identifying and correcting problems with actuators, sensors, and hardware/software, which are often classified into qualitative and quantitative analysis categories. Build a bank of diagnostic observers using structural residual vectors and the syndrome matrix, and then a bank of sliding mode observers using lower-order models of the original system that are invariant to disturbances.

4.3 PROPOSED METHODOLOGY

A Reinforced Kernel-based Artificial Neural Network (RK-ANN) with maximum safety is proposed in this article to identify unexpected problems in AUVs. To begin, the state-sensor data are collected as datasets for this inquiry and normalized during the preparation step to remove any irrelevant information. We next demonstrate how these patterns may be paired with operator-supplied semantic labels to detect AUV defects and diagnose them using an RK-ANN classifier utilizing a natural autoregressive-based topic modeling method. RK-ANN significantly decreases the time-consuming load, simplifies the diagnostic method, and enhances efficiency when compared to traditional model-based diagnosis for AUV safety management. In the actual world, the effectiveness of RK-ANN was shown on a tiny quadrotor AUV known as "Haizhe." The schematic representation of the suggested method is depicted in Fig. 4.3.

4.3.1 Sensor Data

First, the "Haizhe" fault type was established and documented. Then, the initializer was run to see whether each component was able to function correctly. After that, "Haizhe" ran the main application and activated the data logging feature. After then, "Haizhe" started to float submerged. When a failure occurs, it is important to note that new behavioral reactions will be created, but the basic program will remain the same. "Haizhe" would rise to the surface once the main program was complete and cease capturing data. A text file would then be saved in the "Haizhe" file system (also called a data sample). "Haizhe" was tested many times for each defect category. Haizhe ran the same software 10–20 times to verify that the state data was lengthy enough each time. When the test results were utilized as a data sample, fault type associated with

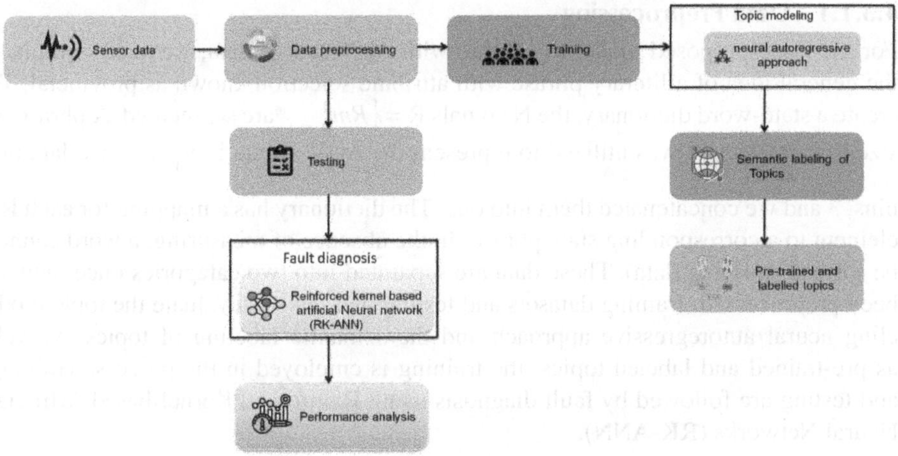

FIGURE 4.3 Schematic representation of the proposed methods.

TABLE 4.1
Data Description

Fault State	Dataset size	Training	Testing
State 1	182	146	36
State 2	260	208	52
State 3	249	199	50
State 4	266	213	53
State 5	268	214	54

that sample became its real label. The combined training and test datasets make up the final dataset, was created by randomly stratifying the data samples into groups. 20% of the whole dataset was devoted to the test data Table 4.1 shows the exact number. To train the model, the training dataset will be utilized, and the test dataset will be used to evaluate its performance.

- Sequence processing: establishing a preset length for string of characters. In cases where original sequence length is inadequate, final instant state data will be repeated; in cases where the original sequence length is excessive, an data's tail will be removed. The sequence length in experiment was set at 196.
- It is possible to determine the average and effective voltages of a motor using the duty ratios of PWM waves.
- The 0–1 normalization technique is used to transfer each feature to the interval [0, 1]. This method can accelerate gradient descent and increase a rate of convergence of a model.

4.3.1.1 Data Preprocessing

For the issue proposed model to operate with AUV sensor data, we need to replace the general idea of a literary phrase with attribute selection known as provincial. To create a state-word dictionary, the N signals $R = \{Rm\}_{m-1}^{m}$ are segmented. A phrase of size $L = \sum_{M-1}^{M} N_K$ was utilized to represent the AUV's state in v_m non-overlapping bins, 3 and we concatenated them into one. The dictionary has a mapping for each R_m element to a corresponding state phrase. In the absence of measuring, a word cannot be formed (missing data). These data are separated into two categories once they've been preprocessed: training datasets and testing datasets. To evaluate the topic modeling neural autoregressive approach and the semantic labeling of topics, as well as pre-trained and labeled topics, the training is employed in the process. Training and testing are followed by fault diagnosis using Reinforced Kernel-based Artificial Neural Networks (RK-ANN).

4.3.1.2 Topic Modeling using Neural Autoregressive Approach (NAA)

We begin with the initial posting NAA. NAA is a generative model over vectors of binary observations $m \in \{0,1\}^F$. Through the probability chain rule, it decomposes $t(m) = \prod_{x=1}^{F} t(m_x \mid m_{<x})$ and computes all $t(m_x \mid m_{<x})$ using the feed-forward architecture

$$l_x(m_{<x}) = \text{sigm}\,(a + E_{:,<x} m_{<x}),\ t(m_x = 1 \mid m_{<x}) = \text{sigm}\,(g_x + m_x, : l_x\,(m_{<x})) \quad (4.1)$$

For $x \in \{1, \ldots, F\}$, where sig $m(i) = 1/(1 + \exp(-i))$, $E \in S^{L\times F}$ and $M \in S^{F\times L}$ are connection parameter matrices, $g \in S^F$ and $a \in S^L$ are bias parameter vectors, $m_{<x}$ is the sub vector $\left[m_1, \ldots, m_{x-1}\right]^P$ and $E_{:,<x}$ is a matrix x – 1 constructed from E's first x columns, rows, and diagonals. Parallel hidden layers $l_x(m_{<x})$ and weighted connections between m_x and each hidden unit $l_{xy}(m_{<x})$ are used in this design.

Equation 4.1 gives all the essential conditions to calculate $t(m) = \prod_x t(m_x \mid m_{<x})$. Stochastic gradient descent may then be used to learn the parameters {g, a, E, M} by mitigating the negative document.

There is a direct connection between the NAA's connectedness (i.e., the existence of a distinct hidden layer $l_x(m_{<x})$ with each $t(m_x = 1 \mid m_{<x})$ weight-sharing and the NAA's. This is a distribution across v derived from an unguided knowledge graph wherein hidden categorical parameters l connect with measurements m via an efficiency functional W(m, l).

$$W(m,l) = -l^P Em - l^P m - a^P l \quad t(m) = \sum_l \exp\left(-W\left(m,l\right)\right) / Y \quad (4.2)$$

In this case the "partition function" (Y) assures that t(m) is a legitimate distribution and that it equals 1. It is difficult to compute the conditional $t(m_x = 1 \mid m_{<x})$ in an RBM, but it may be approximated using mean-field estimations. As $t(m_x, m_{>x} l \mid m_{<x})$ a

product of independent Bernoulli distributions, $(m_u = 1 | m_{<x}) = \mu_u$ (x) and w $(l_x = 1 | m_{<x})$ = τ_y (x), mean-field inference approximates the entire conditional $t(m_x, m_{>x} l \mid m_{<x})$.

$$\tau_y(x) \leftarrow sigm\left(a_y + \sum_{u \geq x} E_{yu\mu u}(x) + \sum_{u<x} E_{yumu}\right), \mu_u(x))$$

$$\leftarrow sigm\left(g_u + \sum_{y} E_{yuPy}(x)\right) \tag{4.3}$$

It is possible to approximate $o(m_x = 1 | m_{<x}) = \mu_x(x)$ using the variational parameter o. An alternative method, NAA, relies on untying the weights between the equations and training the truncation of the output directly to match the available data, rather than using a single application of every message-passing equation. The feed-forward architecture of Equation 4.1 is the outcome.

In the conditionals $t(m_x = 1 | m_{<x})$ sharing the hidden layer parameters is critical, and the NAA does just that. In reality, a lack of inspiration has led to fewer effective outcomes.

4.3.1.3 Semantic Labeling of Topics

When a succession of measurements is conducted on the AUV's present condition, these measurements disclose the mechanisms involved that create such values. So these themes must be connected to the AUV's control algorithm or behaviors and indicate the complex interplay among these risk controls and how effectively it accomplishes their function. To use the topic framework, an AUV's successful analysis may be built directly from the learning algorithm.

The NAA topic modeling approach allows us to learn about an AUV's normal state and specific problems by analyzing a succession of performance patterns. After the training process is finished, a trained topic model is examined, and each of the subjects that were taught is given a semantic meaning. This step must be completed before the trained topic model can be used for categorization. Using an annotated dataset it is feasible to statistically or qualitatively establish the link that a learned topic and a class. We offer a mathematically valid method for determining how closely a topic and a class are linked conceptually.

We calculate a marginal likelihood distribution that decides which subject is relevant for each time step labels are most likely to be found in a given class:

$$Q(X = N \mid \text{class}) = \sum_{t \in T} \frac{Q(X = N \mid T)}{|t_{class}|} \tag{4.4}$$

Where t_{class} the class is index and $Q(X = N \mid T)$ is the topic label dispersion for every time interval. We then inverse Q(X=N| class) using Bayesian' theorem and calculate the likelihood function.

$$Q(class\ X = N) = \frac{Q(X = N \mid class)Q(class)}{Q(X = N)} \tag{4.5}$$

The subject label determines the class's chance of being in the class. For example, we may define Q (class) as Q(X=N| class), T represents the overall number of time steps. We then compute be $Q(X = N \mid class)$, using Eq. (4.5).

4.3.1.4 Faults Diagnosis using Reinforced Kernel-Based Artificial Neural Network (RK-ANN)

It is possible to use an empirical pre-trained model Φ on already observed examples of vehicle speed and faults to get a reliable assessment of the vehicle's state from new, undetected data. We check the performance of the system by comparing the dispersion of nation retrieved from each input observation to the dispersion $\phi N \in \Phi$ of learnt range based N " across the given language V. Faulty states may be found by comparing the prevalence of state-words in a particular observation with the prevalence of provincial in an appropriate subject.

To compare the similarities between two distributions, r and s, we utilize the symmetrized Kullback-Leibler (KL) divergence:

$$\text{KL(R, S)} = \frac{1}{2}\,[\text{O(R, S)} + \text{O(S, R)}] \tag{4.6}$$

Where

$$\text{O(R,S)} = \sum_{j=1}^{S} R_j log_{2} \frac{R_j}{S_j} \tag{4.7}$$

To get started with this strategy, it is necessary to have a system in place for a certain application. That's when the real work of preparing or learning begins. When it comes to teaching, there are two main approaches: guided instruction and unsupervised instruction. An instrument for providing the system with needed yields may either be physically "reviewing" the system's execution or providing the requisite yields administered training. There are several sources of information for unattended training; however, they are located inside the system. RK-ANN's architectures rely heavily on the Iterative Learning Process. The unprecedented learning approach is a part of nursing, in which the system is provided data cases (columns) sequentially such that the weights associated with data values remain balanced indefinitely. Additionally, neural systems in RK-ANN can order designs on which they have not been prepared, which is one of the advantages of neural systems in RK-ANN. Mistakes propagate back through the system, dispensing the framework to direct the weights for application to subsequent records. Because RK-ANNs often fluctuate, this strategy is used again and again. These results may be due to a lack of information on which the planned yield is based. A lack of information also prevents systems from working together. As a starting point, the idea is to use a specified cover to convolution with the AUV to provide each fault an incentive as a component in the dark estimates of its nearby diagnosis, and to replace this activity with consistent activities.

The suggested RK-ANN classifier may be explained in full using the following step-by-step information.

Phase 1: Before beginning the program, set the following settings to their default values:

- Values (w) = 0
- Distortions = 0
- Acquisition rate $\propto$ =1
- S (i, j) = <d(i), d(j)> (4.8)

Phase 2: phase 3 to 8 should be repeated until they cease else.
Phase 3: phase 4 through 6 should be repeated for each new training vector i.
Phase 4: Every interface should be activated as regards:

$$i_x = r_x(x = 1\ \text{ton}) \tag{4.9}$$

Phase 5: To use the continuity formula, calculate the total input:

$$j_{in} = b + \sum_{x}^{n} i_x \omega_{xy} + s(i, y) \tag{4.10}$$

There are n input neurons, and b represents the bias.

Phase 6: This output signal may be used to acquire the final outcome for each output unit:

$$y = 1 \text{ to } v - d(j_{in}) = \begin{Bmatrix} 1 \\ 0 \\ -1 \end{Bmatrix} \begin{matrix} \text{if } j_{inj} > \theta \\ \text{if } \text{if} -, \leq j_{inj} \leq , \\ \text{if } j_{inj} < -\theta \end{matrix} \tag{4.11}$$

Phase 7: Modify the weight and bias for i = 1 to n and y = 1 to v as given below

$$\text{Case 1:if } j_y \neq t_y \text{ then,} \tag{4.12}$$

$$E_{xy}(new) = E_{xy}(old) + \propto t_y i_x b(new) = b(old) + \propto t \tag{4.13}$$

Case 1:if $j_y = t_y$ then,

$$E_{xy}(new) = E_{xy}(old) \tag{4.14}$$

$$B(new) = b(old) \tag{4.15}$$

here 'j' is the actual output and 't' is the target output

Phase 8: Preliminary test for the stopping situation that occurs when mass w is held constant.

4.4 PERFORMANCE ANALYSIS

Using a reinforced kernel-based artificial neural network (RK-ANN) with the maximum level of safety, this research proposes a machine learning (ML) technique for diagnosing unanticipated errors in unmanned aerial vehicles (UAVs). As a starting point for this inquiry, state-sensor data are obtained as datasets, and these datasets are normalized during the preparation step to exclude any undesired data. We next demonstrate how successfully these patterns can be paired with operator-supplied semantic labels to detect AUV defects and diagnose them using the RK-ANN classifier, which is based on a natural autoregressive-based topic modeling approach. Comparing RK-ANN to traditional model-based diagnosis in the context of AUV safety management, the former significantly decreases the time-consuming load; the latter simplifies the diagnostic method and the latter increases efficiency. A demonstration of RK-on ANN's tiny quadrotor unmanned aerial vehicle (UAV) dubbed "Haizhe" demonstrated its effectiveness in the real world. The parameters Detection Rate, Accuracy, Detection Time, True Positive Rate (TPR), and False Positive Rate (FPR) are used. The existing methods include Complex Networks (CN) (Zheng, et al.,2019), Workload Balanced Algorithm (WBA) (Chen and Zhu, 2018), Hybrid Optimization Routing Management (HORM) (Robinson et al., 2021), and Principal Component Analysis (PCA) (Jiang et al., 2019).

4.4.1 Detection Rate

Detection rate is the number of hits per pulse, which is generally one every 100–1000 pulses; count rate is the number of species identified in a given period. The percentage of people with a certain ailment who test positive for that condition when using a gold-standard approach.

$$DR = \frac{TP}{TP + FN} * 100 \tag{4.16}$$

TP = True Positive
FN = False Negative

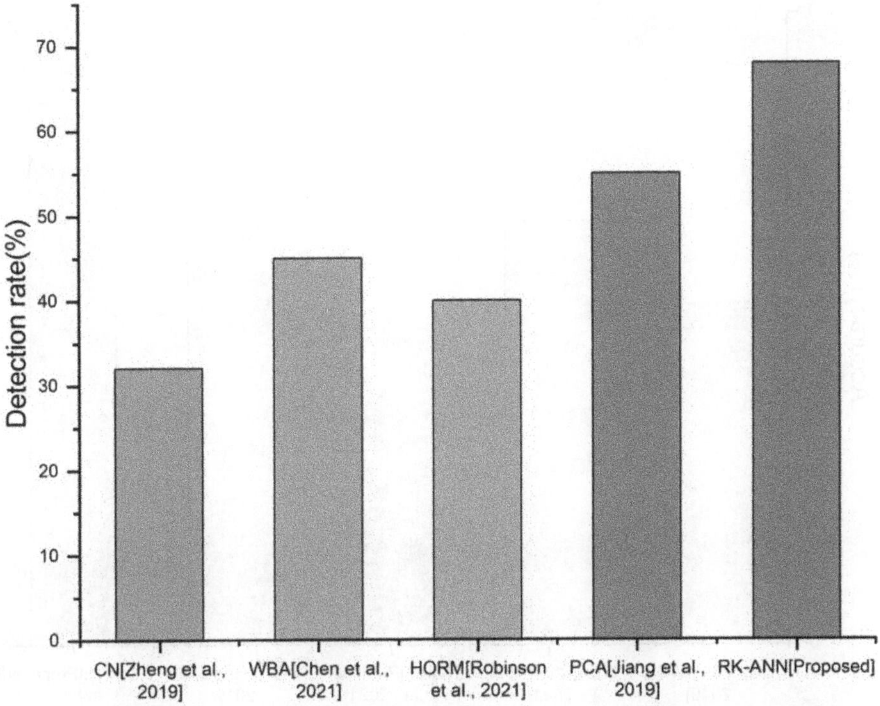

FIGURE 4.4 Detection rate of proposed and existing methods.

It was defined as the number of reference transmitters detected each minute throughout a mission leg. Differences in cruise speed between the AUV and boat meant that completing a mission leg may take longer or shorter depending on the weather conditions. It was necessary to normalize the number of transmitter detections by the amount of time the AUV and boat were within receiving range to make the tracking findings comparable. AUV and boat time were proxies by the period between the first and last record, which began before the transmitters could be detected and ended as detections decreased. Fig. 4.4 depicts the detection rate of proposed and existing methods. It has been shown that the RK-ANN approach has a greater detection rate than previous methods (CN, WBA, HORM, and PCA).

4.4.2 Accuracy

A test's accuracy is measured by its ability to distinguish between healthy and diseased individuals. A test's accuracy may be estimated by looking at the percentage of true positives and false negatives in all of the instances that have been analyzed. It's possible to write this out mathematically as follows:

$$\text{Accuracy} = \text{TP} + \text{TN}\ \text{TP} + \text{TN} + \text{FP} + \text{FN} \tag{4.17}$$

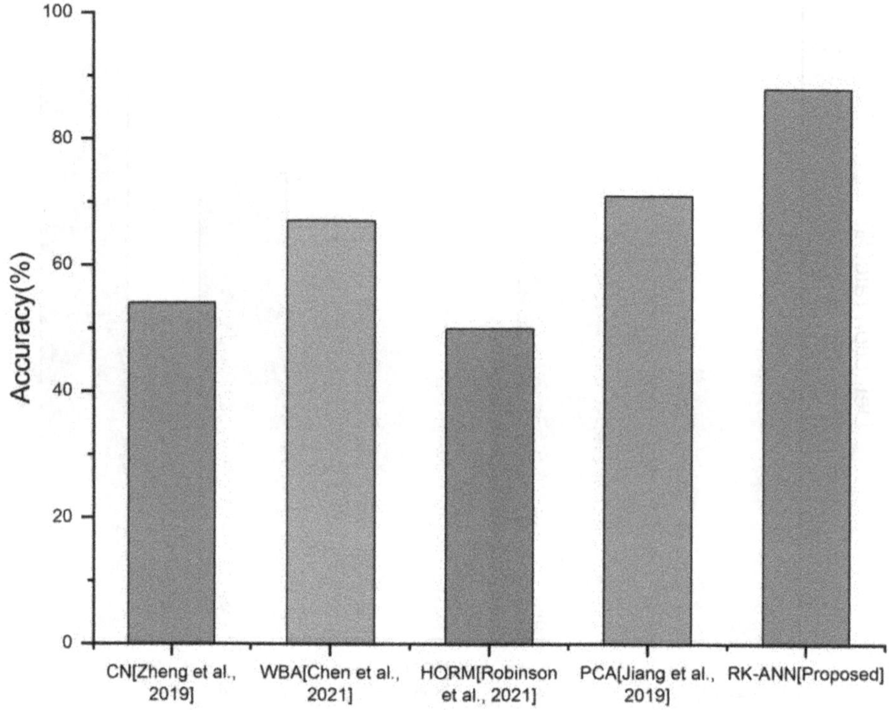

FIGURE 4.5 Accuracy of proposed and existing methods.

The accuracy for fault detection in AUV is shown in Fig.4.5. The accuracy of suggested and current techniques is compared, and as a result, the proposed methods have a higher degree of accuracy when compared to the existing methods (CN with 54 percent, WBA with 67 percent, HORM with 50 percent, and PCA with 71 percent)

4.4.3 Detection Time

When a process takes a long time to complete, the term "detection time" is used. The better the results, the shorter the detection times are known to be in the suggested methodologies.

The proposed and current approaches' detection times are shown in Fig.4.6. When comparing the suggested approaches to the current methods, the new methods have a shorter detection time. The suggested approaches have 45 seconds, followed by the current methods, which are CN with 70 seconds, WBA with 64 seconds, HORM with 58 seconds, and PCA with 72 seconds.

4.4.4 True Positive Rate

The false positive rate (FPR) is the percentage of true positives divided by total false negatives (TP+FN). The probability that a positive will be detected is known as TPR.

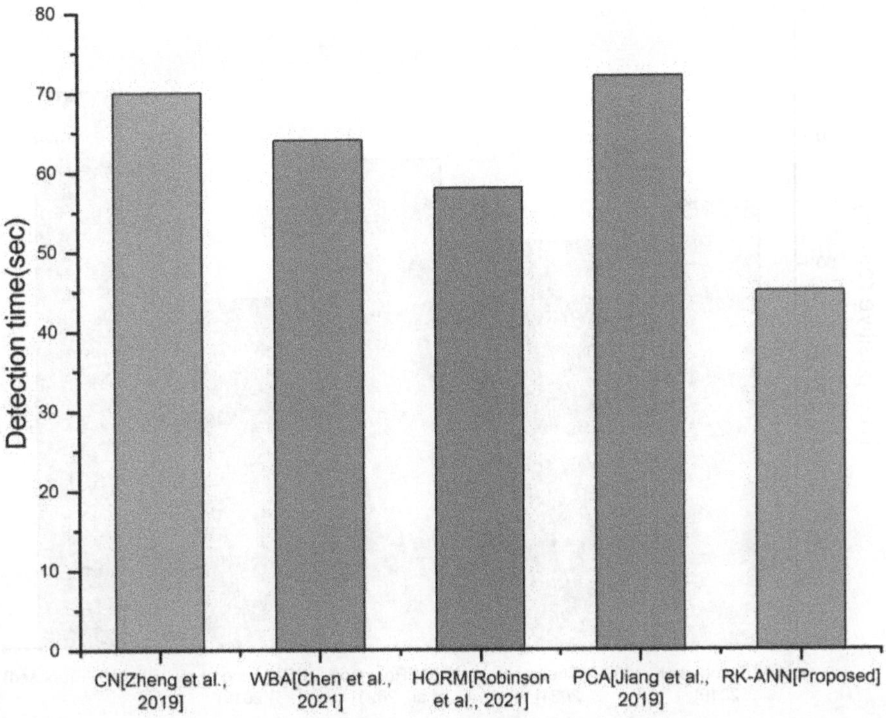

FIGURE 4.6 Detection time of proposed and existing methods.

The likelihood that a real negative will show up as negative on a test is known as the genuine negative rate (also known as specificity).

$$TPR = \frac{TP}{TP} + FN \tag{4.18}$$

The True Positive Rate for proposed and existing methods is indicated in Fig.4.7. When comparing the suggested techniques to the current methods, the TPR in the proposed methods results in a high value. TPR is a suggested approach (RK-ANN) was 88 percent, followed by established methods such as CN, which received 70 percent, WBA, which received 64 percent, HORM, which received 77 percent, and PCA, which received 55 percent.

4.4.5 False Positive Rate

The falsepositive rate is determined as the ratio between the number of negative events that are mistakenly classified as positive (false positives) and the total number of negative events that are correctly classified as negative (true negatives).

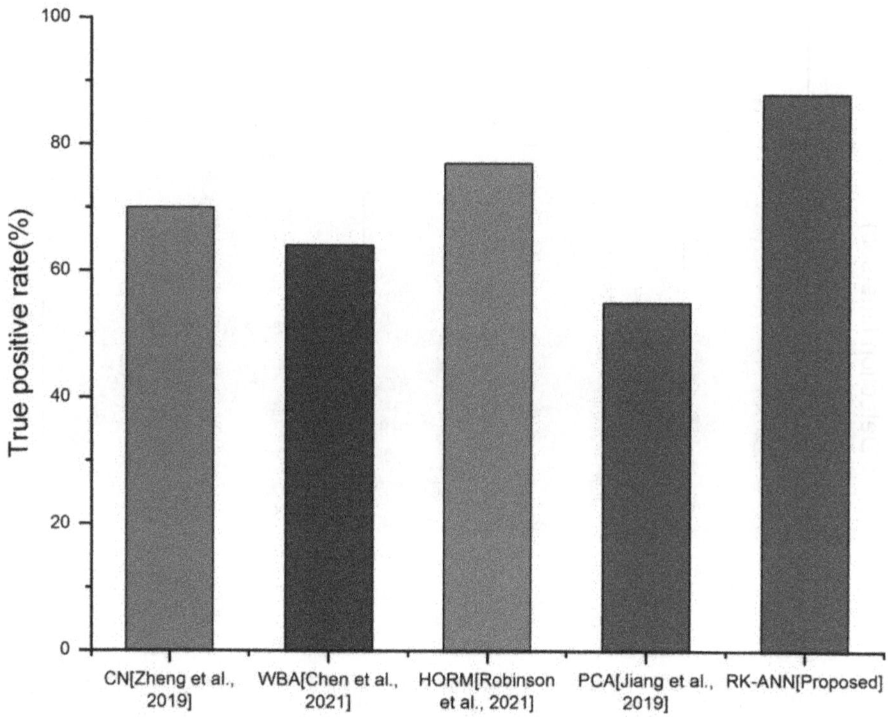

FIGURE 4.7 True positive rate of proposed and existing methods.

$$\text{FPR}=\frac{FP}{FP}+TN \tag{4.19}$$

As seen in Figure 4.8, both the proposed and current approaches have a high rate of false positives. Comparing the proposed approaches to the present methods, the FPR in the proposed techniques has a low value, which is consistent with the current methods. A 40 percent FPR was achieved in recommended ways (RK-ANN), which was followed by proven methods such as CN, which got 65 percent; WBA, which received 65 percent; HORM, which received 50 percent; and PCA, which obtained 70 percent.

4.5 CONCLUSION

An AUV's performance patterns were analyzed using the paradigm given in this paper. We used datasets gathered throughout three consecutive field deployments to train the topic model and assess its classification performance. According to our findings, the framework was able to identify and categorize flaws in the AUV's vertical plane performance with a high likelihood and a low rate of false detection. The system recognizes the relationship between its imposed command strategy and future ride

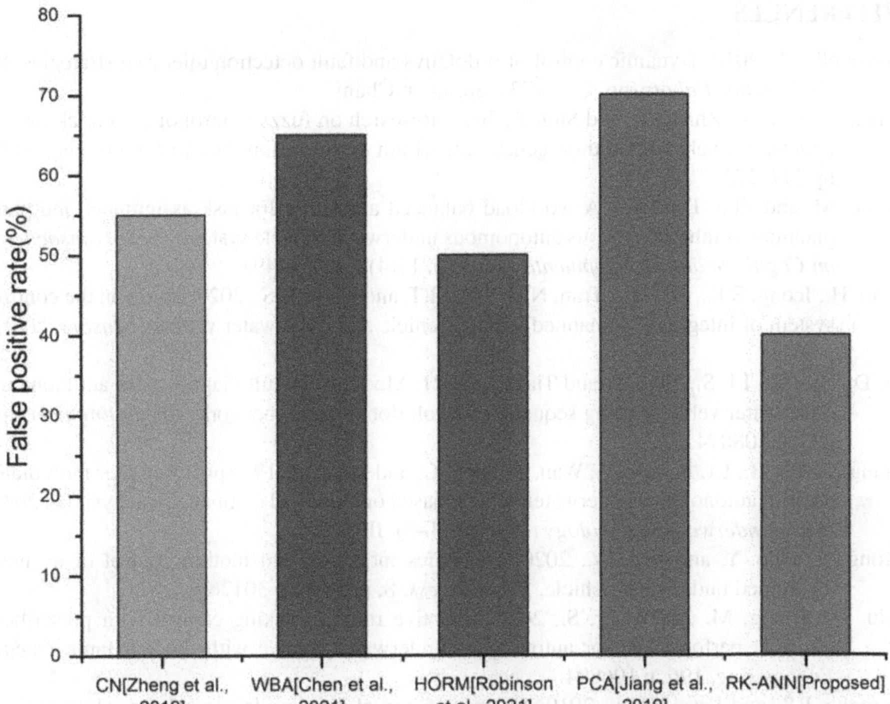

FIGURE 4.8 False positive rate of proposed and existing methods.

quality since it does not rely on professional expertise. Even though the architecture was demonstrated using an AUV in this study, it is relevant to any driverless vehicle. As part of our continuing research, we're evaluating how well the suggested framework performs compared to alternative approaches. One of the goals of our research is to create an architecture for health monitoring that can learn about performance topics live and then use that knowledge of the system's status to automatically replan and choose mitigation measures in response to failures. The suggested approach still needs some improvement. Experimenters may never be able to build up all potential sorts of faults. There is, however, the possibility of using current life-long learning methods, enabling the system to continue to learn new fault patterns. Because of their complementing merits, we want to study a diagnostic technique that incorporates both model-based and model-free methods. In addition, several autonomous underwater vehicles (AUVs) have recently gained a lot of interest. Leaderless or leader-follower modes are used to combine many unmanned aerial vehicles (AUVs) into one mission. Only a few studies have dealt with faults in numerous AUV systems, even though considerable research efforts have been done on control system design and cooperative control. Because of this, a mechanism for diagnosing faults in many AUV systems would be ideal.

REFERENCES

Antonelli, G., 2018. Dynamic control of 6-dofauvs and fault detection/tolerance strategies. In *Underwater Robots* (pp. 111–173). Springer, Cham.

Chen, J., Zhu, H., Zhang, L. and Sun, Y., 2018. Research on fuzzy control of path tracking for underwater vehicle based on genetic algorithm optimization. *Ocean Engineering,* 156, pp.217–223.

Chen, M. and Zhu, D., 2018. A workload balanced algorithm for task assignment and path planning of inhomogeneous autonomous underwater vehicle system. *IEEE Transactions on Cognitive and Developmental Systems*, 11(4), pp.483–493.

Cho, H., Jeong, S.K., Ji, D.H., Tran, N.H., Vu, M.T. and Choi, H.S., 2020. Study on the control system of integrated unmanned surface vehicle and underwater vehicle. *Sensors*, 20(9), p.2633.

Ji, D., Yao, X., Li, S., Tang, Y. and Tian, Y., 2021. Model-free fault diagnosis for autonomous underwater vehicles using sequence convolutional neural network. *Ocean Engineering,* 232, p.108874.

Jiang, Y., He, B., Lv, P., Guo, J., Wan, J., Feng, C. and Yu, F., 2019, April. Actuator fault diagnosis in autonomous underwater vehicle based on principal component analysis. In *2019 IEEE Underwater Technology (UT)* (pp. 1–5). IEEE.

Kong, F., Guo, Y. and Lyu, W., 2020. Dynamics modeling and motion control of an new unmanned underwater vehicle. *IEEE Access,* 8, pp.30119–30126.

Liu, X., Zhang, M. and Wang, S., 2020. Adaptive region tracking control with prescribed transient performance for autonomous underwater vehicle with thruster fault. *Ocean Engineering,* 196, p.106804.

Londhe, P.S. and Patre, B.M., 2019. Adaptive fuzzy sliding mode control for robust trajectory tracking control of an autonomous underwater vehicle. *Intelligent Service Robotics,* 12(1), pp.87–102.

Nie, X., Fan, T., Wang, B., Li, Z., Shankar, A. and Manickam, A., 2020. Big data analytics and IoT in operation safety management in under water management. *Computer Communications,* 154, pp.188–196.

Rangel, M.A.G., Manzanilla, A., Suarez, A.E.Z., Muñoz, F., Salazar, S. and Lozano, R., 2020. Adaptive non-singular terminal sliding mode control for an unmanned underwater vehicle: Real-time experiments. *International Journal of Control, Automation and Systems*, 18(3), pp.615–628.

Robinson, Y.H., Vimal, S., Julie, E.G., Khari, M., Expósito-Izquierdo, C. and Martínez, J., 2021. Hybrid optimization routing management for autonomous underwater vehicle in the internet of underwater things. *Earth Science Informatics*, 14(1), pp.441–456.

Vu, M.T., Le, T.H., Thanh, H.L.N.N., Huynh, T.T., Van, M., Hoang, Q.D. and Do, T.D., 2021. Robust position control of an over-actuated underwater vehicle under model uncertainties and ocean current effects using dynamic sliding mode surface and optimal allocation control. *Sensors*, 21(3), p.747.

Wang, G., Yang, Y. and Wang, S., 2020. Ocean thermal energy application technologies for unmanned underwater vehicles: A comprehensive review. *Applied Energy,* 278, p.115752.

Wang, Y., Tang, C., Wang, S., Cheng, L., Wang, R., Tan, M. and Hou, Z., 2021. Target tracking control of a biomimetic underwater vehicle through deep reinforcement learning. *IEEE Transactions on Neural Networks and Learning Systems,* 33(8), pp.3741–3752.

Xia, S., Zhou, X., Shi, H., Li, S. and Xu, C., 2021. A fault diagnosis method based on attention mechanism with application in Qianlong-2 autonomous underwater vehicle. *Ocean Engineering*, 233, p.109049.

Yan, Y. and Yu, S., 2018. Sliding mode tracking control of autonomous underwater vehicles with the effect of quantization. *Ocean Engineering*, 151, pp.322–328.

Zhao, Y., Chen, Y., Chen, L., Deng, X., Jiang, N. and Hou, L., 2021, October. Intelligent Fault Diagnosis Method for AUV Integrated Navigation System. In *2021 IEEE International Conference on Unmanned Systems (ICUS)* (pp. 896–901). IEEE.

Zheng, X., Feng, C., Li, T. and He, B., 2019. Analysis of Autonomous Underwater Vehicle (AUV) navigational states based on complex networks. *Ocean Engineering,* 187, p.106141.

5 Internet of Things and Machine Learning for Transportation System Using Adaptive Enhanced K-Nearest Neighbor Algorithm

Shivlal Mewada, K.S. Riya, Brahmadesam Viswanathan Krishna, and G. Arunkumar

5.1 INTRODUCTION

Transportation Systems (TS) can help with logistics and transportation. Internet of Things (IoT) approaches for agility and vehicle movement may have several advantages. For instance, reducing the severity of potential accidents, streamlining vehicle movement, and reducing the amount of waste produced by society (Brincat et al., 2019) and (Mahdavinejad et al., 2018). ML techniques are being used in logistics, where autonomous vehicles may move items in safe, monitored infrastructures, choosing the optimal routes for shipments and items to reduce product turnaround schedule. These use cases include road traffic solutions that count, categorize, and identify autos using powerful ML algorithms employing cloud-based techniques and sensors mounted on the road exterior to decrease congestion and accidents (Al-Turjman and Lemayian, 2019).

Network surveillance, which incorporates provisions and monitors for humidity, road surface wetness, and other variables, can help with conservative management and occurrence navigation by alerting a driver or driverless vehicle (Zhang and Lu, 2020). Forecasts and recommendations may be made in real-time, considering diverse networks in a homogenous environment, as well as defined measures that, when compared to felt measurements, assure zero measurement error (Nallaperuma et al., 2019). In this scenario, prescriptive maintenance, which consists of a trained monitoring ML network and identifies abnormalities to recognize the circumstances

DOI: 10.1201/9781003320074-5

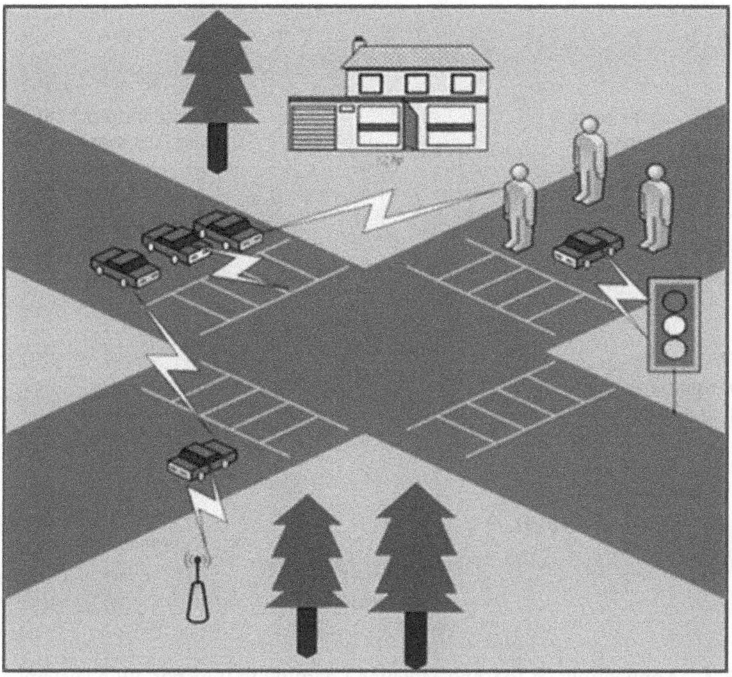

FIGURE 5.1 General schematic diagram of IoT and Machine Learning-based transportation system.

under which an anomaly could occur, is becoming a growing industry for smart cities (Stergiou et al., 2020). In these types of algorithms, ML plays a crucial role since it represents the natural progression of predictive algorithms. To optimize algorithms and approaches, several tasks should be planned (Younan et al., 2020; Din et al., 2019). Figure 5.1 shows the general schematic diagram of IoT and ML-based TS.

Node administration, Monitoring, data-driven programmable rules, data storage and analysis, and IoT systems have additional functions (Priyanka et al., 2018). According to the program, it can be essential for analyzing specific information in IoT devices as opposed to a centrally managed node, as is the case with "cloud computing" architecture (Jan et al., 2019). Developing computer paradigm is recognized as "edge computing" arises when computation moves partially to end network nodes (Vadhwani and Thakor, 2021). Using sensors inserted in cars, portable devices, and city infrastructure, it is possible to give the best route ideas, quick parking inquiries, expensive highway essential services, telemetry for governmental transit, accident prevention, and automatic driving (Tahsien et al., 2020). Figure 5.2 shows the applications of IoT and ML in TS.

As the number of automobiles in major cities grows by the day, it is more important than ever to identify or forecast accidents. One of the most difficult duties

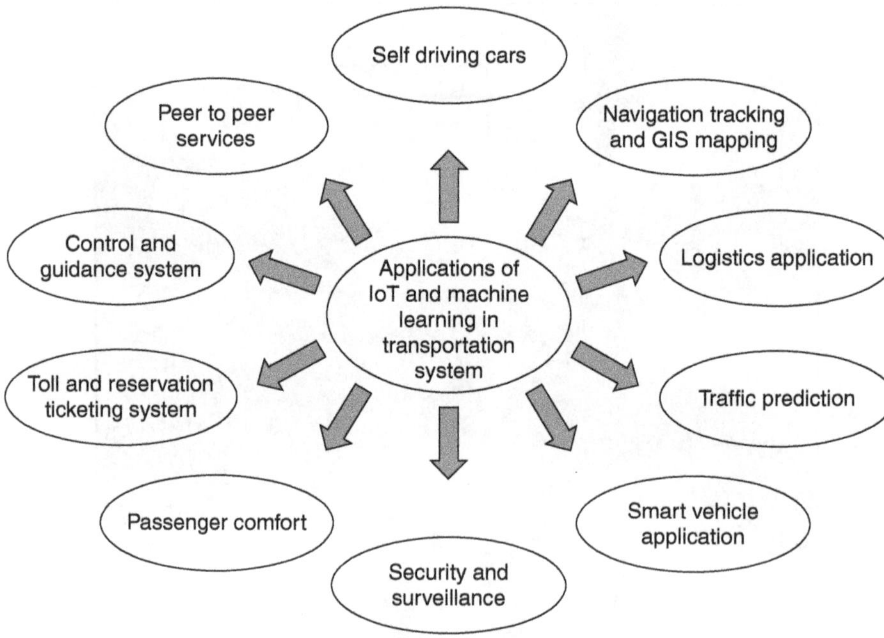

FIGURE 5.2 Applications of IoT and Machine Learning in transportation systems.

in the transportation system is detecting the road surface (Zikria et al., 2019). To identify the vehicle's various features, the authors employed cell phone sensors and ML algorithms. In TS, real-time position monitoring, as well as current status information like speed and driving behavior, is critical (Klaib et al., 2021). In this paper, the authors discuss several statistical ML methods and secure architectures for accident prediction in the TS. The utilization of a traffic light system and data collecting is also employed to forecast urban regions. Our purpose in this chapter is to suggest that computational intelligence be used in a large data set to anticipate road accidents. To forecast the accident in TS, many researchers used ML techniques. To identify accidents in TS, certain strategies such as crowdsourcing data from real-time traffic accidents, black spot detection, and time prediction of road occurrences have been proposed.

5.2 PROBLEM STATEMENT

A cybercriminal attack, a poor security management strategy, a breakdown of IT security features, or harm inflicted by disgruntled workers are all examples of cyber risks. Financial loss, reputational injury, and identity theft can all occur if data in information systems are not effectively safeguarded. Software device makers may be held accountable if they fail to meet IoT performance expectations if their devices fail to adhere to appropriate data policies.

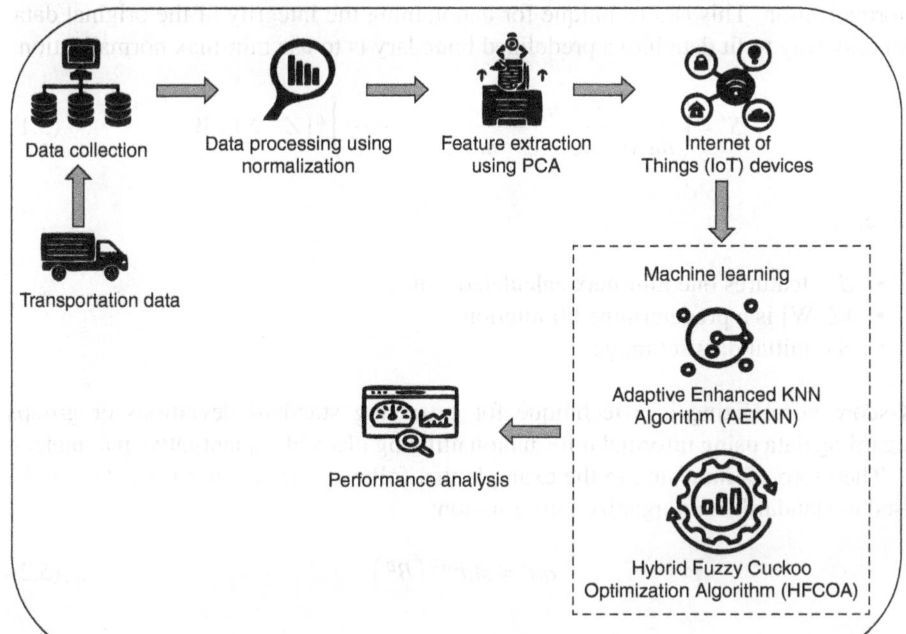

FIGURE 5.3 Illustration of the proposed technique in a schematic.

5.3 PROPOSED WORK

This chapter introduces a novel approach to resolve the problems and improve the performance of the transportation system. In this chapter IoT and ML based AEKNN and HFCOA are used for the transportation system. Figure 5.3 shows a schematic illustration of the suggested technique.

5.3.1 Data Collection

We gathered data on the transportation system, including traffic volume counts, traffic speed statistics, bicycle or pedestrian counts, and parking occupancy in this chapter.

5.3.1.1 Data Preprocessing Using Normalization

Preprocessing involves preparing and transforming data so that it can be used efficiently. Preprocessing data involves minimizing data size, establishing relationships, normalizing data, removing outliers, and extracting data characteristics.

Several strategies are employed, such as cleansing, integrating, transforming, and minimization.

Normalization is primarily concerned with reducing or eliminating redundant data. Taking a set of data and performing linear transformations on it is known as Min-Mix

Normalization. This is a technique for maintaining the integrity of the original data. An easy way to fit data into a predefined boundary is to use min-max normalization.

$$S' = \left(\frac{S - minvalue\,of\,S}{maxvalue\,of\,S - min\,value\,of\,S} \right) * (Z - S) + W \tag{5.1}$$

Here,

- S, - features one min-max calculated value,
- [Z, W] is a predetermined limitation
- S – initial data set range

K-score normalizing is a technique for extracting standard deviations or groups regarding data using informal information utilizing ideas like quantitative parameters.

Therefore, as indicated in the example that follows, the K-score variable may be used to standardize unorganized information:

$$qa' = std^{qa-}\left(B^{\bar{B}}\right) \tag{5.2}$$

- q_a' - K score adjusted value of one.
- q_a - value of a row $\bar{B}$

$$std^{(B)} = \sqrt{\frac{1}{(r-1)} \sum_{j=1}^{r} \left(qa - \bar{B}\right)^2} \tag{5.3}$$

$$\bar{B} = \frac{1}{r} \sum_{j=1}^{r} qa \tag{5.4}$$

In decimal scaling, a scale is established from -1 to 1. Therefore, decimal scaling is used

$$q^a = qx \tag{5.5}$$

- q^a - scaled value
- q - range of values x–smallest integer Max(|q^a|)<1

A linear conversion of the original data is achieved through min-max normalization.

Assume that the lowest and highest possibilities for parameter E are Min_x and Max_x.

By computing: [new-min_x, new-max_x], through min-max normalization, e´ in the range is transferred from E to e´.

$$E' = ((e - min_x)/(max_x - min_x)) * (new - max_x - new - min_x) + new - min_x \tag{5.6}$$

In w-score normalization, the data for parameter E are normalized using the average score of E. Formula for normalizing E to e′ is as follows:

$$e' = \left(\left(e - \bar{X}\right) / \sigma_X\right) \tag{5.7}$$

where the sample variance of parameter e is e and σX, respectively. It is advantageous to use this normalizing approach if true range of e is a mystery, as are its lowest and highest values.

When using decimal scaling for normalization as well as the quantity of the parameter e will have its decimal point shifted to the right. e′ is the absolute maximum value that determines how many decimal points are shifted. When comparing an E value e to a e′ applying this equation:

$$e' = \left(e / 10^i\right) \tag{5.8}$$

In this case, i is the smallest integer such that Max (|e′|) < 1.

5.3.1.2 Feature Extraction Using PCA

Principal Component Analysis (PCA) is often used as a feature extraction method in data analysis. PCA works by finding the covariance matrix's highest-Eigen valued eigenvectors and afterward mapping the information upon a new domain of the same or fewer variables. PCA converts an n-feature matrix into a new dataset with (hopefully) less than n features in practice. To put it another way, it decreases the number of characteristics by developing a separate, quicker amount of indicators that gain a large percentage of the knowledge in the innovative parts. The notion of PCA has been explained very effectively elsewhere, thus the focus of this presentation will be on showing how PCA works in practice.

5.3.2 Internet of Things (IoT)

An IoT device is used to store the acquired transportation data. It can also show data and treat a wide range of connection methods. A brand-new technological application framework called the IOT enables rapid distant information acquisition using contemporary wireless networking and communication systems. The term "Internet of Things" (IoT) refers to the "Internet via all connections"; in other words, data is gathered by putting data into a device that detects information, such as a radio-frequency identification reader. Omnipresent Internet connections enable the acquisition and transmission of a network that intelligently recognizes ubiquitous information. To accomplish a common objective, communicate with one another, and work together, the technological usage designs relies on ubiquitous gathering data instruments such as mobile phones, sensors, tags, drives, and radio-frequency identification (RFID). The IoT is composed of three distinct layers of infrastructure. For the purpose of facilitating the transfer of information, the next level, also known as the layer of transport, provides utilizes preexisting networks such as local and broad-area networks, the global web, and communication systems, as well as Electronic Data

Interchange (EDI), Electronic Product Codes (EPC), and additional analysis of information and transfer advances in technology; The obtained sensor data is processed and used at the third tier, the programming level; Mobile phones and devices, such as personal computers, cellphones, and additional mobile phones, all run operating systems on which applications are delivered and utilized in accordance with organizational strategy. The organizational framework is implemented via the delivery and use of apps on cellular phones and other devices including desktops and laptops, smartphones, and extra cell phones. Increased intelligence and better management of utilities are only a few instances of how cutting-edge technologies like computing in the cloud, social media, and data sharing of resources have been utilized to better the lives of city dwellers. Smart city initiatives are increasingly influencing urban landscapes and economies.

5.3.3 Machine Learning

To refine and construct advanced TS, ML methodologies can be applied. Its major goal is to find models in a set of data and make independent decisions established on an individual occasion. In supplement to separate driving, ML techniques are frequently utilized in associated vehicles, which are cars operated by people that connect to a network of sensors to communicate information. One of the ML techniques is predictive maintenance, which is an extension of forecasting algorithms, or algorithms that strive to reveal problems or anticipate certain activities. Progressive maintenance is capable of not only anticipating but also establishing defined operations autonomously in the case of an occurrence or anomaly. To refine and construct advanced TS, ML methodologies can be applied.

5.3.4 Adaptive Enhanced K-Nearest Neighbor Algorithm (AEKNN)

The Adaptive Enhanced K-Nearest Neighbor approach (AEKNN) evaluates information gain for each attribute and changes the Euclidean distance computation depending on the weighted entropy of the attribute value.

We begin by calculating the entropy of each characteristic. The value of the jth attribute B_j of y is represented by the attribute vector $<b_1,b_2,\ldots\ldots b_n>$, where ai specifies the value of the instance. If |S| is the number of training models and R is a collection of training models, then R's data volume is F(R).

$$F(R) = -\sum M_j log_2\left(M_j\right) \tag{5.9}$$

where $M_j = \left|D_j, R\right| / \left|R\right|$

M_j is a class D variable in R that represents the likelihood of an arbitrary tuple.

It is possible to distribute the training model set into p subsets using B if it has distinct standards. The restrictive information of classifications separated by B may then be calculated using $F\left(R\middle|B\right)$.

$$F(R|B) = \Sigma(|R_j| / |R|) * F(R_j) \quad (5.10)$$

Entropy is used to calculate the information gain of characteristics. The difference between the information in the entire dataset and the data in an element is known as information gain.

$$Gain(B) = F(R) - F(R|B) \geq 0 \quad (5.11)$$

It is crucial to select the k value before the KNN process can begin. A cross-validation technique called leave-one-out is used to compute the k value. There is one training tuple in the validation data, whereas the remaining tuples make up the training data. Each training tuple undergoes this step for authorization purposes.

In the conventional KNN approach, the space between the test place and each training place must be computed. It takes a long time to complete this treatment. Consequently, we separated the whole training set based on similar data positions. The training set is separated into subgroups, which lowers computation time significantly. To calculate the space between the test spot and the data spot, the subset must be divided into test spots and data spots. It is determined which cluster the test point belongs to by the cluster center. In a cluster of data points, a collection facility is calculated as the average.

The space between each training sample in the nearest cluster and the test sample is calculated using the distance formula. The weight of an attribute determines how much information it gains. The formula is as follows:

$$b(Y_i, Y_j) = \sqrt{\Sigma Z_j (b_s(Y_i) - b_s(Y_j))^2} \quad (5.12)$$

where Z_j is the weight that each dataset attribute has been allocated. After that, the distances are categorized, and the k-nearest neighbors are established. The votes of K-Nearest Neighbor are then weighted differently depending on their distance from the test demand. The bulk of the class brands in the chosen cluster is from the test sample. A test case's distance from the test case is inversely proportional to its weight.

$$Z = 1 / (b)^2 \quad (5.13)$$

As a result, the following is the weighted class probability estimate formula:

$$d(y) = arg\max_{d \in D} \sum_{j=1}^{l} z\delta(d, d(X_j)) \quad (5.14)$$

where x1, x2, …, xk are the test instance's K-Nearest Neighbor, k is the number of neighbors, D represents the $\delta(d, d(X_j)) = 1$ if $d = d(y_i)$ and $\delta(d, d(X_j)) = 0$ otherwise.

5.3.5 Hybrid Fuzzy Cuckoo Optimization Algorithm (HFCOA)

The HFCOA is a new optimization metaheuristic algorithm established on the necessary offspring parasitism of some cuckoo species, in which they lay their eggs in host bird nests. Female parasitic cuckoos have developed the capacity to mimic the colors and patterns of a rare different host class's eggs. To ensure that their eggs will emerge first, cuckoos try to find a nest recently laid by another bird, such as a bird that just laid eggs. Other cuckoo chicks can replicate the call of a cuckoo chick. If a cuckoo egg is found in the nest, the present birds either remove it or abandon it. In this optimization strategy, each nest signifies a possible explanation. Following three ideal principles is the easiest approach to employing the HFCOA. There is a guaranteed supply of suitable nesting sites, and the host bird will expose the cuckoo's egg with a predetermined probabilities; the best nests will produce the best offspring, and these offspring will be used in the following generations of the breeding program. Equation 5.15 is used to perform the global random walk.

$$Y_{j\ new} = Z.Y_j + stepsize.randn \tag{5.15}$$

where Z is the mass of inertia, This is calculated using an equation.

$$Z = Z_{max} - \left(\left(Z_{max} - Z_{min}\right) / \, iter_{max}\right).iter \tag{5.16}$$

The following equation was used to get the needed step size value:

$$Stepsize = 0.01.step.\left(Y_j - Y_{best}\right) \tag{5.17}$$

The difference factor $(Y_j - Y_{best})$ indicates that the solution remains unaltered when it is the best answer. Considering the measurement scaling of cuckoo movements while searching for a new place to nest is k, and the step size of walks or flights is K/100, we may assume that the step size will increase by a factor of 0.01. The step factor may be calculated using the following equation:

$$step = \left(\frac{\sigma(\beta).randn}{randn}\right)^{\frac{1}{\beta}} \tag{5.18}$$

where randn is a random number between 0 and 1 randn € [0, 1], where b is a fixed value $(1 \le \beta \le 3)$.

5.4 PERFORMANCE ANALYSIS

This chapter explains a novel strategy for addressing problems with the current transportation system. The HFCOA is used in this chapter to examine the IoT and ML-based transportation system. We evaluated the system's effectiveness in terms of speed control, trip time reduction, accident avoidance, and efficiency. These parameters are compared to the Decision Tree Algorithm (DTA) (Zeadally and Tsikerdekis, 2020), Parallel Genetic Algorithm (PGA) (Abbasi et al., 2020), and Deep Learning (DL) (Lv et al., 2020), which are all known approaches.

5.4.1 Speed Control

The transportation system's most important component is speed control. Using an IR sensor, a speed control-based IoT system may calculate the speed of a vehicle based on time and distance. When compared to existing approaches, the suggested system's speed control performs better. The comparison study for speed control is shown in Figure 5.4.

5.4.2 Travel Time Improvement

Improvements in travel time are linked to increased safety, freight transportation efficiency, and a better quality of life for road users who experience less wait and irritation. The IoT, which combines the three biggest technological and commercial developments at the moment—mobility, automation, and data analytics—has the potential to completely change the way TS acquire data and information. When compared to other current solutions, the suggested strategy is more effective in reducing trip time. The comparison study for trip time improvement is shown in Figure 5.5.

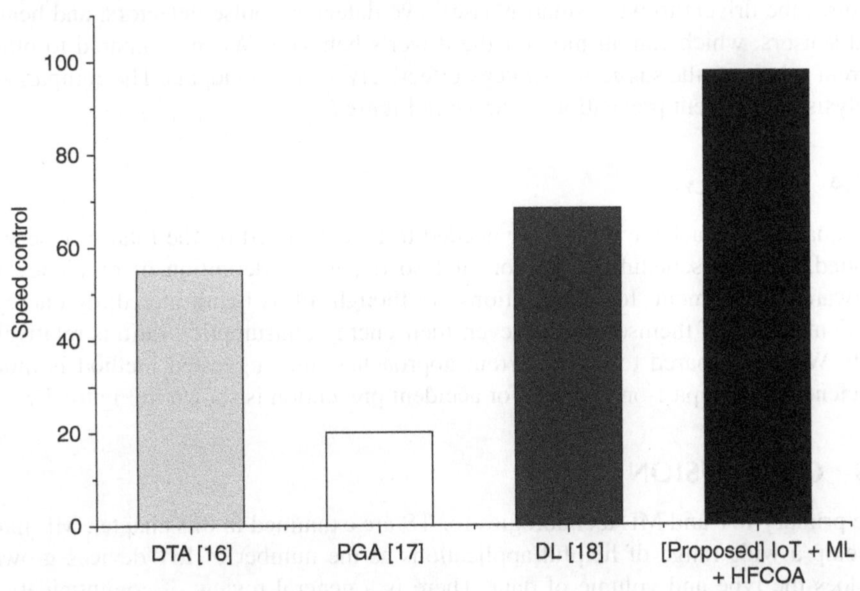

FIGURE 5.4 Comparison of speed control.

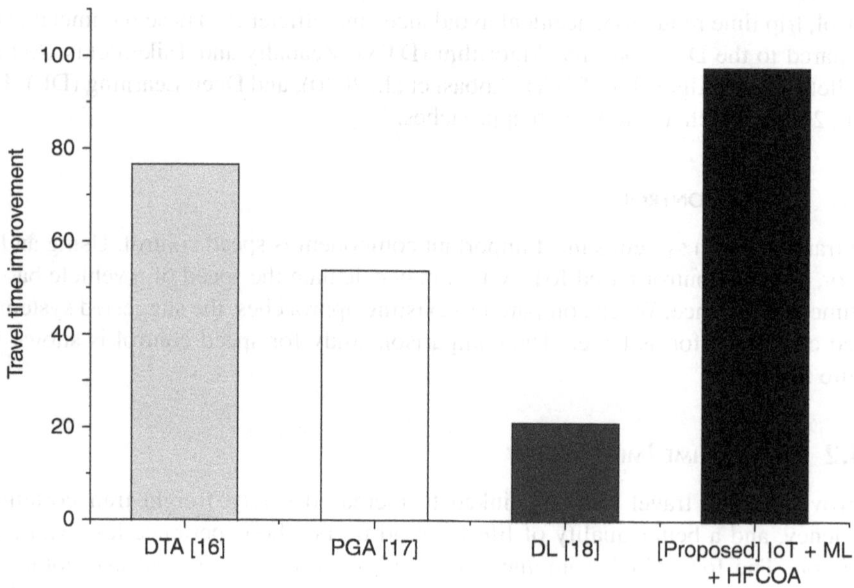

FIGURE 5.5 Comparison of travel time improvement.

5.4.3 Accident Prevention

Vehicles equipped with sensors can identify the potential hazard and share information about their speed and position with other vehicles. IoT-based accident prevention requires the drivers to wear smart glasses, eye detectors, pulse detectors, and heartbeat sensors, which can all monitor the driver's behavior. When compared to other current solutions, the suggested strategy effectively avoids accidents. The comparison analysis for accident prevention is shown in Figure 5.6.

5.4.4 Efficiency

The quantity of package utilized or needed to be calculated by the total of reserves applied, such as schedule, endeavor, and so on, is the definition of efficiency in software development. IoT applications are thought of as being incredibly energy-efficient in and of themselves, however, their energy consumption ratio is relatively high. When compared to other current approaches, the suggested method is more efficient. The comparison analysis for accident prevention is shown in Figure 5.7.

5.5 CONCLUSION

The primary IoT and ML technologies for TS are examined in this chapter. ML may develop a wide range of helpful applications as the number of IoT devices grows, as does the type and volume of data. There is a general review of communication protocols and cutting-edge ML technologies that are crucial in this subject. Finally, more research is carried out in real-world circumstances. The suggested HFCOA

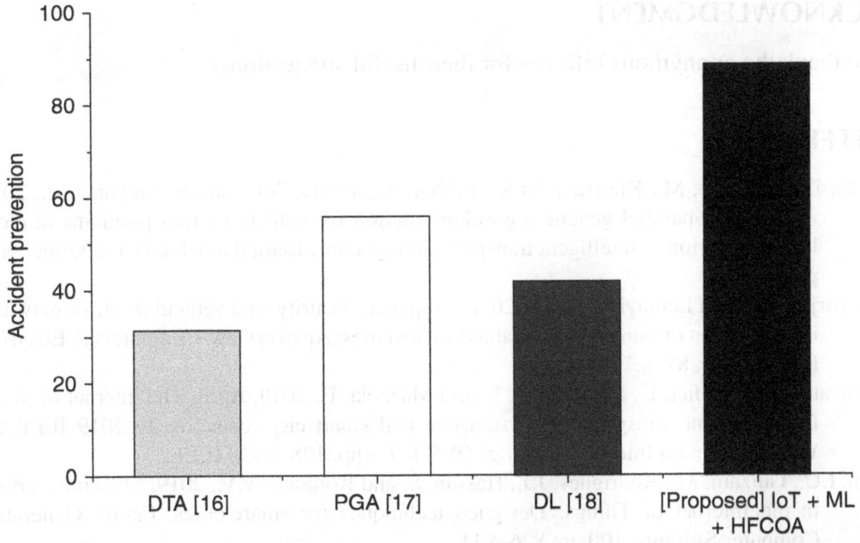

FIGURE 5.6 Comparison of accident prevention.

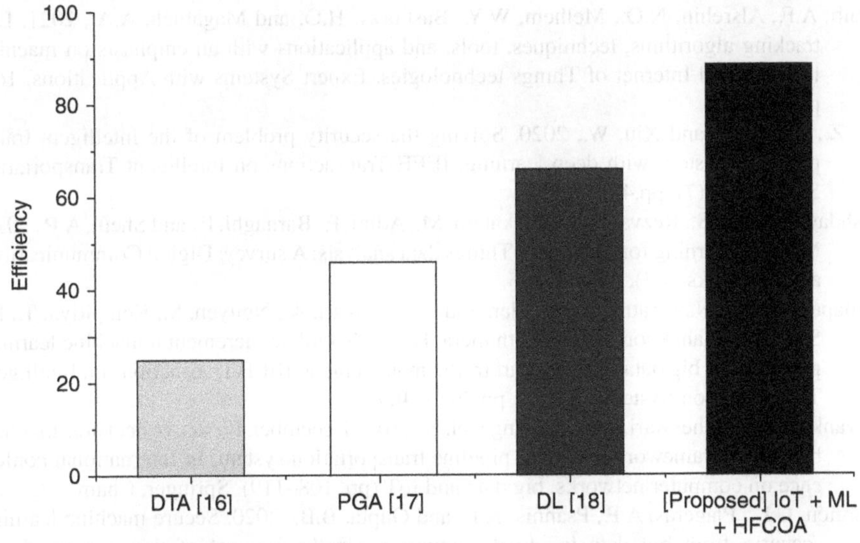

FIGURE 5.7 Comparison of efficiency.

improves the performance of smart lighting systems, parking applications, speed control, and travel time reduction. We compared the suggested approach to various strategies that are currently in use. The suggested transportation system efficiently coordinates the flow of people, products, and vehicles so that routes may be utilized. Through appropriate timetabling and route management, the suggested TS aim to cut transportation costs and enhance delivery times.

ACKNOWLEDGMENT

We thank the anonymous referees for their useful suggestions.

REFERENCES

Abbasi, M., Rafiee, M., Khosravi, M.R., Jolfaei, A., Menon, V.G. and Koushyar, J.M., 2020. An efficient parallel genetic algorithm solution for vehicle routing problems in cloud implementation of intelligent transportation systems. Journal of Cloud Computing, 9(1), pp.1–14.

Al-Turjman, F. and Lemayian, J.P., 2020. Intelligence, security, and vehicular sensor networks in the internet of things (IoT)-enabled smart-cities: An overview. Computers & Electrical Engineering, 87, p.106776.

Brincat, A.A., Pacifici, F., Martinaglia, S. and Mazzola, F., 2019, April. The internet of things for intelligent transportation systems in real smart city scenarios. In 2019 IEEE 5th World Forum on Internet of Things (WF-IoT) (pp. 128–132). IEEE.

Din, I.U., Guizani, M., Rodrigues, J.J., Hassan, S. and Korotaev, V.V., 2019. Machine learning in the Internet of Things: Designed techniques for smart cities. Future Generation Computer Systems, 100, pp.826–843.

Jan, B., Farman, H., Khan, M., Talha, M. and Din, I.U., 2019. Designing a smart transportation system: an internet of things and big data approach. IEEE Wireless Communications, 26(4), pp.73–79.

Klaib, A.F., Alsrehin, N.O., Melhem, W.Y., Bashtawi, H.O. and Magableh, A.A., 2021. Eye tracking algorithms, techniques, tools, and applications with an emphasis on machine learning and Internet of Things technologies. Expert Systems with Applications, 166, p.114037.

Lv, Z., Zhang, S. and Xiu, W., 2020. Solving the security problem of the intelligent transportation system with deep learning. IEEE Transactions on Intelligent Transportation Systems, 22(7), pp.4281–4290.

Mahdavinejad, M.S., Rezvan, M., Barekatain, M., Adibi, P., Barnaghi, P., and Sheth, A.P., 2018. Machine learning for Internet of Things data analysis: A survey. Digital Communications and Networks, 4(3), pp.161–175.

Nallaperuma, D., Nawaratne, R., Bandaragoda, T., Adikari, A., Nguyen, S., Kempitiya, T., De Silva, D., Alahakoon, D. and Pothuhera, D., 2019. Online incremental machine learning platform for big data-driven smart traffic management. IEEE Transactions on Intelligent Transportation Systems, 20(12), pp.4679–4690.

Priyanka, E.B., Maheswari, C. and Thangavel, S., 2018, December. Proactive decision-making-based IoT framework for an oil pipeline transportation system. In International conference on computer networks, big data and IoT (pp. 108–119). Springer, Cham.

Stergiou, C.L., Plageras, A.P., Psannis, K.E. and Gupta, B.B., 2020. Secure machine learning scenario from big data in cloud computing via the internet of things network. In Handbook of computer networks and cyber security (pp. 525–554). Springer, Cham.

Tahsien, S.M., Karimipour, H. and Spachos, P., 2020. Machine learning-based solutions for the security of Internet of Things (IoT): A survey. Journal of Network and Computer Applications, 161, p.102630.

Vadhwani, D. and Thakor, D., 2021. Statistical analysis of vehicle detection in the ITS application for monitoring the traffic and road accident using the internet of things. In Advances in VLSI and Embedded Systems (pp. 55–70). Springer, Singapore.

Younan, M., Houssein, E.H., Elhoseny, M. and Ali, A.A., 2020. Challenges and recommended technologies for the industrial internet of things: A comprehensive review. Measurement, 151, p.107198.

Zeadally, S. and Tsikerdekis, M., 2020. Securing the Internet of Things (IoT) with machine learning. International Journal of Communication Systems, 33(1), p.e4169.

Zhang, H. and Lu, X., 2020. Vehicle communication network in intelligent transportation system based on internet of things. Computer Communications, 160, pp.799–806.

Zikria, Y.B., Kim, S.W., Hahm, O., Afzal, M.K. and Aalsalem, M.Y., 2019. Internet of Things (IoT) operating systems management: opportunities, challenges, and solution. Sensors, 19(8), p.1793.

6 Optimization of Routing Protocols in ITS Using Elephant Herd Algorithm for Underwater Wireless Sensor Networks

M. Vedhapriya and J. Dhilipan

6.1 INTRODUCTION

The use of autonomous underwater vehicles (AUVs) for long-term ocean surveys or other technological tasks is expanding. An AUV can be seen as the tool for producing appropriate and goal-directed behaviour. The "Goal-directed" refers to a robot's conduct that is focused on accomplishing the tasks that the mission specifies. Several payload sensors, including as cameras and sonars, are frequently fitted to autonomous underwater vehicles (AUVs) for the purpose of seeing and examining the undersea environment.

A number of marine and maritime uncrewed systems platforms, such as USVs (uncrewed surface vessels), AUVs (autonomous underwater vehicles), and ROVs, may be equipped with AI (artificial intelligence) software (remotely operated vehicles). These cars have the ability to automatically assess sensor data and learn how to react in extremely dynamic conditions thanks to the power of AI software [1, 2].

IoT nodes are described as being qualified for dependable transit by established routes, despite having heterogeneous memory or effective sensors/actuators with limited memory. Memory restrictions have an impact on link dependability and increase end-to-end mistakes.

In IoT-based networks, communication routes involving these more mobile nodes offer security risks. Self-organization of IoT [3] nodes creates difficulties for application security. Environmental dynamism and random node placement are two drawbacks of WSNs.

 DOI: 10.1201/9781003320074-6

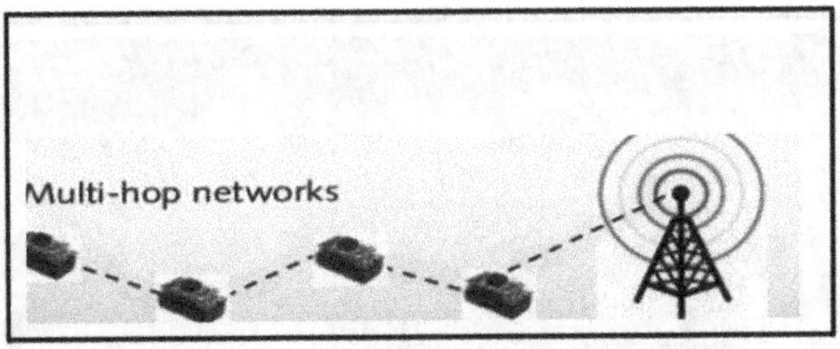

FIGURE 6.1 Multi-hop routing.

Power and processing capacity limitations, which also restrict their operating ranges, lead to unstable communications. MCRPs depend on priority distributions to choose routes, which can be computationally demanding and unreliable and are typically ignored. It is clear from the aforementioned issues that MCRPs neglect to take into account the status of nodes, particularly their lifetime or congestion.

Enhancing network lives and meeting the QoS requirements in MCRPs is a very difficult topic, and a network should have enough energy and routing information to communicate between a source and a destination. An energy effectual and an idyllic multi-path routing practice based on EHO, EHO-ETQRP, for IoT-based WSN applications.

A network service's QoS has historically been described as a set of traffic characteristics. Both performance-oriented and non-performance-oriented criteria may be present in these traits. The non-performance-oriented group establishes the criteria for precedence, price and service level. These are concerned with issues that are linked but do not directly affect how communications perform. Traditional QoS [4] criteria offer a fairly separate perspective on service parameters, which limits their ability to reflect the total QoS offered to the user/application.

6.2 PROTOCOLS BASED ON MULTI-HOP ROUTING

Ad hoc wireless network management is the domain of multi-hop routing technologies. Different strategies have been used in an effort to successfully address the routing issues present in ad hoc networks [5]. These protocols are divided into proactive protocols, reactive protocols, and hybrid protocols, which are three separate groups. The two subcategories of flat and hierarchical routing protocols can be added to the three main categories. Each of these groups and subgroups has benefits and drawbacks of its own.

Network topologies, particularly dynamic topologies, are more beneficial to IoT operations than multimedia communications [6]. This suggests that there is a need for extremely effective MCRPs that address multimedia communications and IoT applications. Any MCRP will not support QoS in multimedia communications [7,8] because its main goal is to maximize [9] resource utilizations while lowering network energy consumptions.

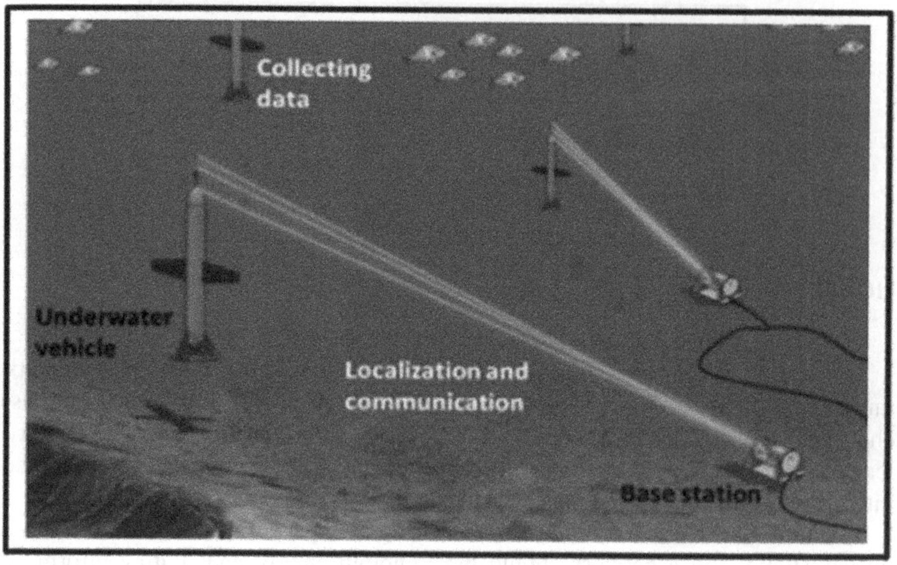

FIGURE 6.2 AUV communication.

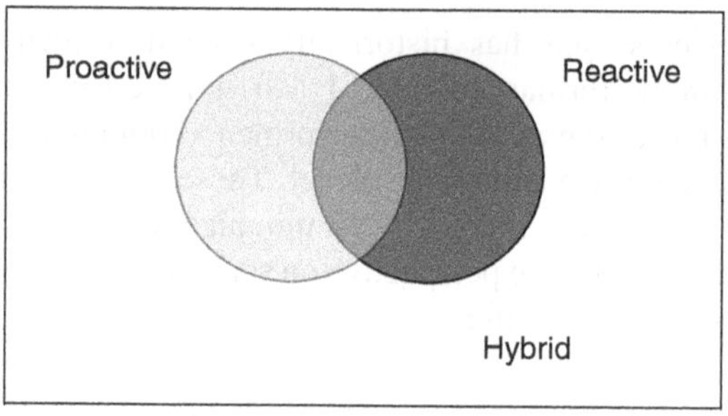

FIGURE 6.3 Autonomous underwater vehicle.

6.3 A PROACTIVE, REACTIVE, OR MIXED STRATEGY

By routinely updating their awareness of the network, proactive routing systems make an effort to always have a clear picture of the network architecture. Because the network's paths have already been set, one benefit of the proactive protocols is that, there are from little to no inactivity when an application initiate transmitting data. Additionally, as directions are revised frequently, they frequently remain current. In big networks, proactive protocols fail because the regular updates consume a significant amount of network resources by creating a lot of control traffic overhead.

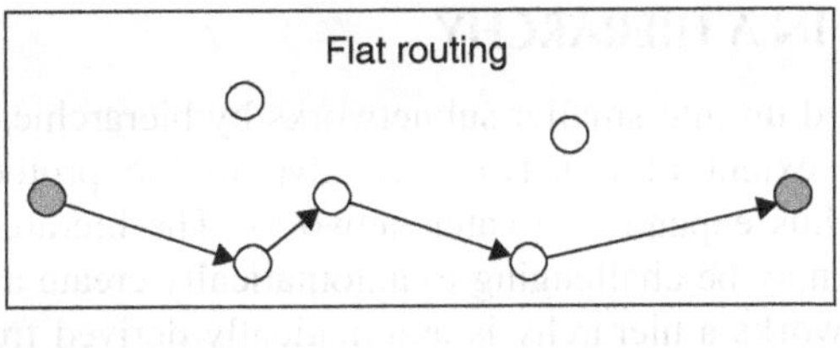

FIGURE 6.4 Flat route protocol.

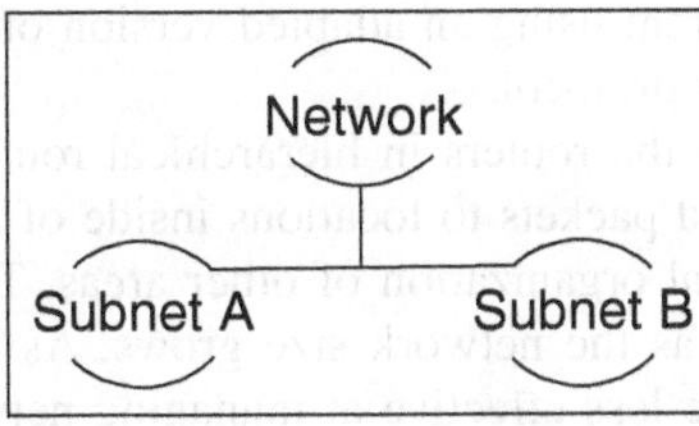

FIGURE 6.5 Hierarchical routing.

6.4 HYBRID PROTOCOLS

Wireless networks (like cellular networks) and wireless sensor networks are both parts of hybrid wireless sensor networks. This network is crucial for overcoming the transmission range [10] and data rate restrictions of conventional sensor networks.

In an effort to find the ideal routing solution, hybrid protocols integrate the finest features of proactive and reactive protocols. Hybrid protocols often take a proactive stance inside a network sector or zone and set-up routes across the zones in a reactive manner. This method has the advantage of that the nodes inside a sector can be contacted rapidly, while the nodes beyond the sector or a zone can be touched without clogging up the network with unnecessary regulator traffic. This method has the drawback that routes amid nodes in different sectors or zones are frequently not ideal.

6.5 FLAT ROUTE

The management of these networks is made easier by a flat route protocol, which treat a network as a single entity while treat all other nodes equally [11]. These protocols struggle to gage to big networks due to their flat routing architecture. Router optimization for link state is an improvement of the traditional link state algorithm designed specifically for mobile wireless LANs which is the optimized link state routing (OLSR) protocol. OLSR uses a selection of nodes referred to as multipoint relays to controllably flood a network with information to route information (MPR). During the flooding process, only MPR nodes are permitted to broadcast messages, offering a significant optimization over conventional flooding [12]. To reduce control traffic overhead, OLSR restricts link status information generation to MPR nodes only. Based on dispersed partial link state data, OLSR determines routes for the network.

6.6 ROUTING IN A HIERARCHY

Networks are divided up into smaller subnetworks by hierarchical routing methods, which reduces the extent of a network relative to the protocol. The hierarchical protocols can thus expand to greater networks. The hierarchical protocols are complicated, and it may be challenging to automatically create and maintain a hierarchy. Buy one networks a hierarchy is automatically derived from the network by the order one network protocol, and the hierarchy is then organized in a tree-like manner. Each node [13] chooses the node in the tree hierarchy that has the most routes through it, as its parent using an adapted version of the (DSDV) Destination Sequenced Distance Vector protocol.

Regions are created for the routers in hierarchical routing. Each router has full instructions on how to send packets to locations inside of its own region. However, it is unaware of the internal organization of other areas. The quantity of routers in the network will increase as the network size grows. As a result, as routing table size grows, routers become less effective at managing network traffic. Hierarchical routing is what we're employing to get around this issue. Routers are categorized in regions, which are groups, in hierarchical routing. Each router contains data about the routers in its own region but none of the other regions' routers are known to it. In their database, routers thus store one entry for each additional region. On the bases of hierarchy, the elephant herd algorithm is derived.

6.7 ELEPHANT COMPORTMENT

Wide elephants are gregarious creatures, and there are various clans within the elephant species. Male elephants remain unsocial and will leave their family group as they mature. The elephants of various clans coexist in groups under the direction of a matriarch [14]. A novel type of swarm-based on heuristic search approach, named EHO, which is proposed for addressing global optimization challenges. It was inspired by the herding behaviour of elephant groups. Clan updating operator and separation operator are idealized versions of the natural behaviour of elephant herding. According on its current location and the matriarch, clan, each elephant in EHO similarly uses a coterie informing operator to update its position. the Elephant Herding Optimization (EHO) algorithm, which was inspired by the social behaviours of elephant herds seen in the wild. Although elephants in the wild display significantly more complex behaviours, the following idealized guidelines were taken into account when creating the EHO algorithm: The elephant population is divided into several clans, each of which contains an equal number of elephants. One elephant in each tribe guides the other members in their hunt for water and food sources. In addition, some population members who reach a particular age in each generation leave their clan to live alone in other areas far from the family to which they belong.

6.8 ELEPHANT HERDING OPTIMIZATION (EHO) ALGORITHM

The performance of EHO is examined through a number of experiments to find substantially more suitable solutions to the majority of benchmark problems [15].

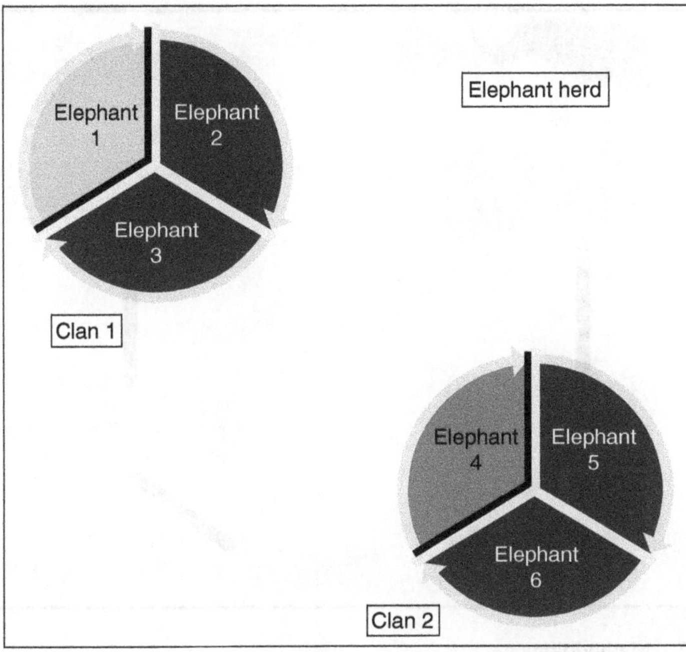

FIGURE 6.6 Elephant herding optimization.

Principles in order to make it global optimization problems three important rules must be followed:

- Within the elephant population, there are various clans, and each clan has a specific number of elephants.
- Every generation with a predetermined number of male elephants will disperse from their family group and live alone in an area separate from the main elephant herd.
- A matriarch oversees the daily affairs of each clan of elephants

Elephants are sociable animals that live in social groups made up of females and their young. Many elephants make up an elephant clan, which is led by a matriarch [16]. Male members typically reside abroad, while female members choose to live with relatives. Up until they entirely separate from their families, they will gradually become independent of them.

(1) Few clans with specific numbers of elephants make up the majority of the elephant population.
(2) Every generation, a predetermined number of male elephants will disperse from their family group and live alone in an area separate from the main elephant herd.
(3) Each clan's matriarch commands the elephants.

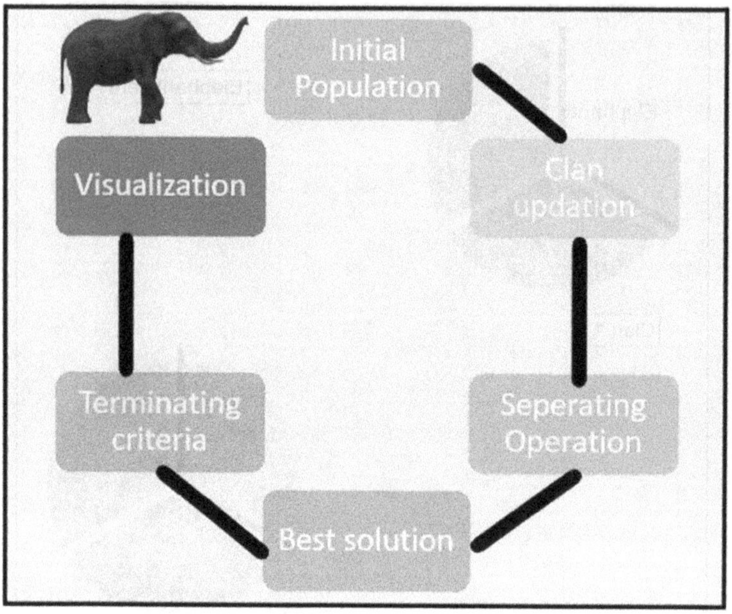

FIGURE 6.7 Methodology.

Operator for Clan-updating – Elephants in each clan are led by a matriarch in accordance with their normal behaviours. As a result, matriarch ci has an impact on each elephant's new position. Equation can be used to calculate the elephant j in clan ci (6.1).

$$xnew,\ c1,\ 2{=}xc1,\ 2 + a \times (xfinest,\ c1 - xc1,\ 2) \times r \tag{6.1}$$

where $x_{new,c12}$ and x_{c12} that the current and the new position, previous position for elephant 2 in clan $c1$, respectively. The $x_{finest,c1}$ is matriarch c_1 which represents the best of the elephant in the clan. $a \in [0,1]$ indicates, a scale factor, $r \in [0,1]$. The finest elephant can be calculated by Equation (II) for each clan.

$$xnew,c1,2 = \beta \times xcenter,c1 \tag{6.2}$$

Where the $\beta \in [0,1]$ denotes a factor which controls the impact of the $x_{center,c1}$ on $x_{new,c1,2}$. $x_{new,c1,2}$ is the novel individual. $x_{center,1i}$ is the midpoint of the individual clan *ci*. It can be intended by Equation (III) for the yth dimension.

$$xcenter,c1,d = 1nci \times \Sigma j = 1nc1xc1,2,y \tag{6.3}$$

Where the values of $1 \leq$ (is lesser) $d \leq$ (lesser) D and n_{c1} show the number of elephants in clan *ci*. x_{c12y},representing the *y*-th dimension of elephant discrete x_{c12}. $x_{center,c1}$ is the middle of clan *ci* and it can be restructured by Equation (6.3).

When resolving optimization issues, the separating process through which male elephants depart from their family group can be modelled as a parting operator. The elephant individual with the worst fitness in each generation implements the separation operator.

Equation (6.4).

$$xvilest,c1 = xMin + (xMax - xMin+1) \times rand \tag{6.4}$$

where the maximum value of x_{Max} represents the higher bound of the individual and x_{Min} shows the lesser bound of the individual. $X_{vilest,c1}$ indicates the worst individual in clan c_1. *rand* [0, 1] is a stochastic distribution between 0 and 1.

6.9 FLOW CHART FOR EHO

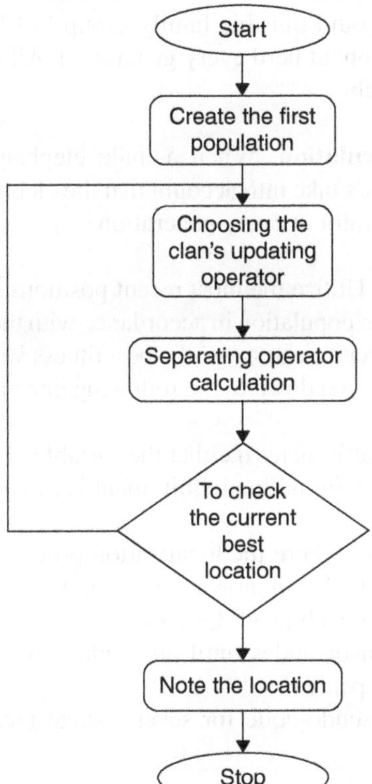

EHO Algorithm Steps

1) Creating the First Population
2) Choosing the clan's updating operator
3) Separating operator calculation
4) Note down the current best location/ optimal current remedy
5) Stop values

Creating the first population: In the group of mammals, elephants are the biggest creatures on earth. Elephants are larger, more docile, and have different behavioural patterns. A female elephant with her calves or other closely related female elephants are examples of complicated groups in which an elephant resides [11]. These organizations are known as clans. Elephants can range in size from three to twenty-four in a clan. Matriarch refers to a clan's head elephant herd in concentric circles, led by matriarch.

Choosing the clan updating operator: Elephant herding behaviour is used to address a global optimization problem. These guidelines help us to streamline the issue. after eliminating the population's poorest values [17]. The clan updating operator is used to update the fittest values, and the worst values are ignored. It is believed that there are two groupings, such as clans, that make up the entire elephant population. There are a specific number of elephants in these clans. It is also said that the elephant male who performs the worst would quit his family group [18,19]. It lives by itself at a great distance from the elephant herd every generation. All elephants who dwell in a clan are led by the matriarch.

Separating operator calculation: When a male elephant reaches adulthood, he leaves his family group. Let's take into account that the elephant with the worst fitness will use the separation operator at every generation.

Optimal current remedy: Utilize the most recent positions to assess the population's fitness values. Sort the entire population in accordance with the new rules for feasibility, and then save the best answer as x best and the best fitness value as y min. The top elephant, elephant x best, is passed down to the following generation as elephant x0.

Stop value: The algorithm terminates if either the variable cycle ranges the determined number of iterations or the termination requirement is satisfied.

Several stage localizations, where the localization process is carried out in phases, for EHO implementation. Only unknown sensor nodes with three or more nearby anchor nodes are located at each level. One stage's localized sensor node transform into the stage's anchor sensor nodes until all unidentified nodes are localized, the localization procedure is repeated.

Method 2 provides a pseudo-code for several stage localization using the EHO algorithm.

1 Initialize the network topology for several stage localization using EHO algorithm.
2 Track down undiscovered nodes that have three or more anchor nodes nearby.
3 Use the EHO technique to locate the unidentified nodes that were discovered in step 2.
4 Enhance localized sensor nodes in the existing stage to the anchor node list and delete them from the unknown node list.
5 Repeat steps 2–4 until all sensor nodes are localized.

6.10 CONCLUSION

To handle the various combinatorial optimization problems or in the engineering optimization field, population-based optimization algorithms are extremely popular and frequently employed. Inorder, to surge the overall endurance of the network system, this effort is implementing Elephant Herd Optimization (EHO), which is similar to cluster head formation in underwater sensor networks. The work is based on dynamic clustering, with the dynamic cluster head chosen because it resembles an elephant tribe. The improvements in artificial intelligence allow for the integration of deep learning, a highly optimized approach, to enhance this task. Additionally, soft computing strategies and paradigms drawn from nature can be combined to achieve higher degrees of accuracy and optimization. These methods could include things like the dynamics of river creation, the bat algorithm, the cuckoo search, the lion optimization, and many others. The EHO algorithms were enhanced by researchers, who effectively used them in a variety for optimization of the domains. Each AUV on a set of points that are dynamically produced to assemble the whole group close to the current physical field source location. If there are multiple sources, the trajectory is built to identify the one that is responsible for the physical field's greatest value. However, there are a number of components of the EHO approach that need more research.

REFERENCES

[1] TCP for mobile ad hoc networks: J. Liu, and S. Singh . Atcp, *IEEE Journal on Selected Areas in Communications*, 2001.

[2] Ad hoc on-demand distance vector (aodv)routing: E. M. Belding-Royer, and C. Perkins, IETF Internet Draft, 2003.

[3] Real-time price-based demand response management for residential appliances via stochastic optimization and robust optimization: Z. Chen, L. Wu, and Y. Fu, *IEEE Transactions on Smart Grid,* 3(4), pp. 1822–1831., 2012.

[4] Wireless sensor networks: a survey on recent developments and potential synergies: P. Rawat, K. D. Singh, H. Chaouchi, and J. M. Bonnin, *Journal of Supercomputing*, vol. 68, pp. 1–48, April 2014.

[5] Wireless sensor network localization based on cuckoo search algorithm: S. Goyal and M. S. Patterh, *Wireless Personal Communication*, vol. 79, pp. 223–234, May 2014.

[6] Elephant herding optimization: G.-G. Wang, S. Deb, and L. dos S. Coelho, in *Proceedings of the 2015 3rd International Symposium on Computational and Business Intelligence (ISCBI)*, pp. 1–5, December 2015.

[7] Improvement of the wireless sensor network lifetime using LEACH with vice-cluster head: A. Mehmood, J. L. Mauri, M. Noman, and H. Song, *Ad Hoc & Sensor Wireless Networks*. Aug;28 (1–2), pp. 1–7, 2015.

[8] Broadcasting over duty-cycled radio using LEACH clustering for delay/power efficient dissimilation in wireless sensor networks: M. Khiati and D. Djenouri. BOD-LEACH, *International Journal of Communication Systems*. Jan 25;28(2), pp. 296–308, 2015.

[9] Differential-evolution control parameter optimization for unmanned aerial vehicle path planning: K. Kok, and P. Rajendran, *PLOS ONE*, 11(3), p.e0150558, 2016.

[10] A new metaheuristic optimisation algorithm motivated by elephant herding behaviour: G.-G. Wang, S. Deb, X.-Z. Gao, and L. dos S. Coelho, *International Journal of Bio-Inspired Computation,* vol. 8, pp. 394–409, January 2017.

[11] Performance of elephant herding optimization algorithm on CEC 2013 real parameter single objective optimization: V. Tuba, M. Beko, and M. Tuba, *WSEAS Transactions on Systems*, vol. 16, pp. 100–105, 2017.

[12] Distributed algorithm for target localization in wireless sensor networks using rss and aoa measurements: S. Tomic, M. Beko, R. Dinis, and P. Montezuma, *Pervasive and Mobile Computing*, vol. 37, pp. 63–77, June 2017.

[13] Support vector machine optimized by elephant herding algorithm for erythemato-squamous diseases detection: E. Tuba, I. Ribic, R. Capor-Hrosik, and M. Tuba, *Procedia Computer Science*, 122, pp. 916–923, 2017.

[14] An efficient hybrid of elephant herding optimization and cultural algorithm for optimal design of trusses: Jafari, M., Salajegheh, E. and Salajegheh, J., *Engineering with Computers*, 35(3), pp. 781–801, 2018.

[15] Energy efficient data gathering in wireless sensor networks using rough fuzzy C-means and ACO: S. Mondal, S. Ghosh, and P. Dutta, in *Industry Interactive Innovations in Science, Engineering and Technology* (pp. 163–172). Springer, Singapore., 2018.

[16] Balancing trade off between data security and energy model for wireless sensor network: B. E. Manjunath, and P. V. Rao, *International Journal of Electrical and Computer Engineering (IJECE)*, 8(2), 1048–1055, 2018.

[17] Hybrid meta-heuristic optimization based energy efficient protocol for wireless sensor networks: S. Kaur, and R. Mahajan, *Egyptian Informatics Journal*, 2018.

[18] Enhanced elephant herding optimization for global optimization: A. Ismaeel , I. Elshaarawy, E. Houssein, F. Ismail, and A. Hassanien, *IEEE Access*, 7, pp. 34738–34752, 2019.

[19] https://transpireonline.blog/2019/08/19/new-nature-inspired-metaheuristic-algorithm-for-elephants-elephant-herding-optimization-eho-algorithm/

7 Effect of the Intelligent Transportation System into the Sustainable Environment and Ecology System

Kshatrapal Singh, Ashish Kumar, Manoj Kumar Gupta, and Ashish Seth

7.1 INTRODUCTION

Transportation has grown critical to societies and the economic growth of day-to-day operations. However, relying on motorized transportation as a daily activity contributes significantly to global climatic changes. Given that fossil fuels account for 93% of transportation energy, separating transportation growth from emissions increase appears unlikely without key policy or technology advancements [1–2].

It's natural that there hasn't been consensus on a comprehensive approach to decreasing the greenhouse emissions of transportation because there are so many options. Continuing to support less carbon technological innovation and deployment; trying to encourage modal shifting from private vehicle use to less pollution choices like walking, cycling, as well as public transportation; trying to advocate more effective transport governance [3] and driving actions; and shift the focus that seek to eliminate such a need to travel altogether are some of the main emissions reduction pathways. The major points of contention appear to be behavioral and technological innovation, as well as reform vs dramatic reform.

While cognitive behavioral efforts to discourage private vehicle usage would result in considerable reductions in transportation-related emissions, a technologically reformer solution could be perceived as being more socially and politically acceptable. A transformation to a lower-carbon lifestyle will very probably need improvements in and across all behavioral and technical aspects, as well as reformist and revolutionary approaches [4–6]. From a socio-technical standpoint, the various components—actors, artefacts, and institutions—must be linked in a certain sequence for a sustainable transition to a reduced pollution option to occur.

DOI: 10.1201/9781003320074-7

Furthermore, since transportation is such a challenging industry to decarbonize, much focus has been placed on piecemeal technological advancement to provide highly sustainable transportation choices in the past. As a result, intelligent technologies are becoming increasingly crucial in the effort for sustainability practices; yet socio-behavioral constraints may still have to be overcome if these innovations are to be widely used.

7.2 INTELLIGENT TRANSPORT SYSTEM AN OVERVIEW

In order to achieve a less carbon future, ICT serves as a cross-sectorial facilitator in 2 situations. Whether in systems that exist along the transportation sector (such examples, in healthcare, educational, business, or related non-transport services) or in organizations that are fully integrated with sections of the transportation mechanism (like as parts of the vehicle, road infrastructure or management approach). In practice, rather than being mutually exclusive, these two situations can overlap [7–9]. A description of ICT's role in both situations is presented below, along with how each represents technological advancement or fosters behavioral change, which is a crucial step in building the case for ITS as a long-term option.

Much more has indeed been stated about the advent of ICT, the supported information networks, as well as the societal and environmental impacts of machine learning. Significantly, this ICT growth should be directed with respect to less-carbon businesses, where the widespread adoption of intelligent approaches may benefit the economy while also accelerating the transition to environmental protection.

When ICT is used as a supplementary to the transportation system, it allows people to make decisions that affect both for the consumer spending for transportation and the requirement for particular modes, such as allowing people to work from homes or promoting mode switching. Where ICT is integrally embedded in the transportation mechanism [10–13], which helps in optimizing the usage of the infrastructure in certain direction, whether it's for effectiveness, safety, or a better travel experience, or even a more sustainable style of functioning. Figure 7.1 represents the ITS architecture.

ICT can help minimize carbon-intensive physical transportation activities when it is used to integrate complementary fields and activities to transportation. This happens when communication system over a digital connection takes the place of physically negotiating geographical space. As a result, ICT can lessen the demand for social and business travel in this setting. Preventing needless travel by connecting to the internet and arranging unavoidable travel utilizing real-time traffic data before accessing the transportation network are both in line only with small choices strategy. Better decisions are a soft approach that use the incorporation of travel plans to positively affect travelers and promote better sustainable discretionary conduct in the schools, business, or homes [14]. Even though the research is inconclusive, it appears that businesses are starting to recognize the importance of addressing environment problems in its travel schemes in order to strengthen their industrial citizenship as well as engage prospective employees.

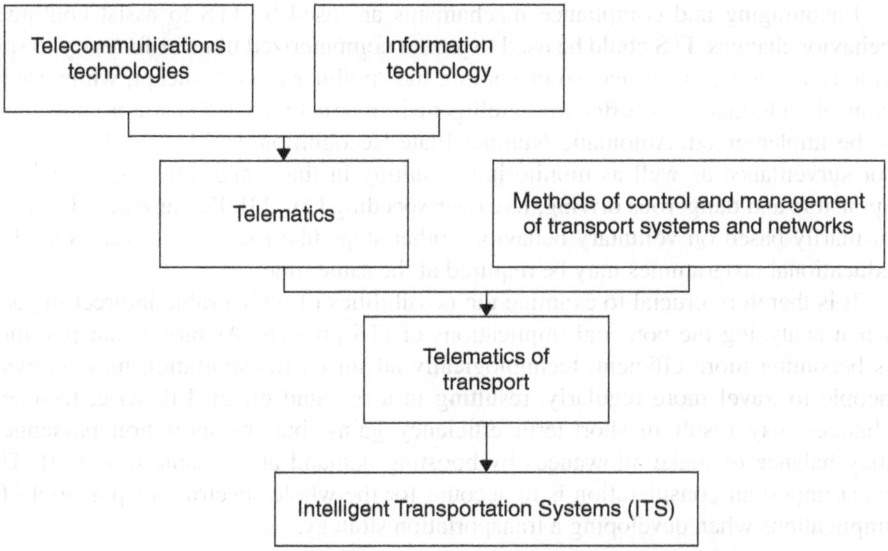

FIGURE 7.1 Intelligent transport system.

When ICT is incorporated into a transportation approach, this is considered to be like an Intelligent Transportation System (ITS). Its architecture, such as variable-message indicators and digital toll collection systems, can be set to work in a specific place. Satellite tracking systems, enhanced traveler systems engineering, and adaptive cruise control are forms of mobile ITS architecture that can be found on platform cars or on people. Cooperative systems are a type of modern ITS, that uses vehicle-to-vehicle communication, vehicle-to-infrastructure communication, and infrastructure-to-vehicle communication to improve transportation planning system by improving road effectiveness and security [15–17, 18].

Incorporated ITS solutions use variable and pervasive connections to detect transportation activity, analyze this in-coming data, and after that convey information from the actual to public/private transportation users and/or traffic controllers, resulting in more efficient transportation networks. Drivers can be given information to help them make decisions while moving through the transportation system, control their behavior, act as warnings, or even make decisions for them if a quick response is needed. This knowledge has the potential to empower individuals to pursue better sustainable actions that lead to a reduced carbon tomorrow.

ITS can impact and advise the driver's role in supporting ecological driving or eco-driving, in addition to actively improving traffic management through information providing. Abnormal and harsh driving movements, irregular driving speed, unexpected halting and going, and unnecessary speeding and idling are all examples of these driving skills. It also refers to proper vehicle repairs in order to maximize efficiency [19–20]. Eco-driving might be encouraged through knowledge, promotion, and regulation with thoughtful legislation and preparation.

Encouraging and compliance mechanisms are used by ITS to assist continued behavior changes. ITS could be used to permit computerized transfer for entry to specific route, zone, or service, so promoting the "polluter pays" concept, while access control techniques can offer the intelligent infrastructure needed for pricing plans to be implemented. Automatic Number Plate Recognition could yield the potential for surveillance as well as monitoring, resulting in fines, and thus mitigating irresponsible and dangerous driving, like over-speeding [21–23]. Because eco-driving is primarily based on voluntary behaviors, other steps like fiscal incentives as well as educational programmes may be required at the same time.

It is therefore crucial to examine the possibilities of unfavorable indirect impacts when analyzing the potential implications of ITS projects. As motor transportation is becoming more efficient, technologically advanced transportation may motivate people to travel more regularly, resulting in a rebound effect. Likewise, technical changes may result in short-term efficiency gains, but transportation passengers may balance or make allowances by boosting demand at the same time [24]. The most important consideration is to account for the whole spectrum of potential ITS implications when developing a transportation strategy.

7.3 THE ADVANTAGES OF ITS

The benefits on application of Intelligent Transportation Systems are as follows:

- Cost of rolling managing inventory are reduced.
- Cost of paving maintenance and repairs are reduced.
- Growth in the region's financial advantages.
- Increased throughput by 25–30% on the street.
- Increased road traffic survival (50–90% reductions in traffic accidents).
- Drivers, passengers, and pedestrians will benefit from improved travel convenience and driving congestion.

7.4 THE PHYSICAL ARCHITECTURE OF ITS

Constructing ITS architectural features, i.e., a sequence of interconnected systems (logical, structural, and transportation) between many components of machines incorporating Intelligent Transportation Systems to increase cost-effective results that are also easily maintained and manageable, are among the most crucial objectives set by countries/regions/cities for themselves because incorporating intelligent services in transportation. ITS can also help with the creation and maintenance of transportation infrastructure, such as railways [25].

Many governments have not specified any particular ITS technology or providers; instead, these are open systems which improve the competitiveness of deployed technologies. These, on either hand, are "isolated" by design, in the sense that they each accomplish their assigned tasks independently, but when joined, they might be contradictory and incapable of sustaining each other. Moreover, data from various ITS places should be combined. Numerous studies have shown that the data gathered

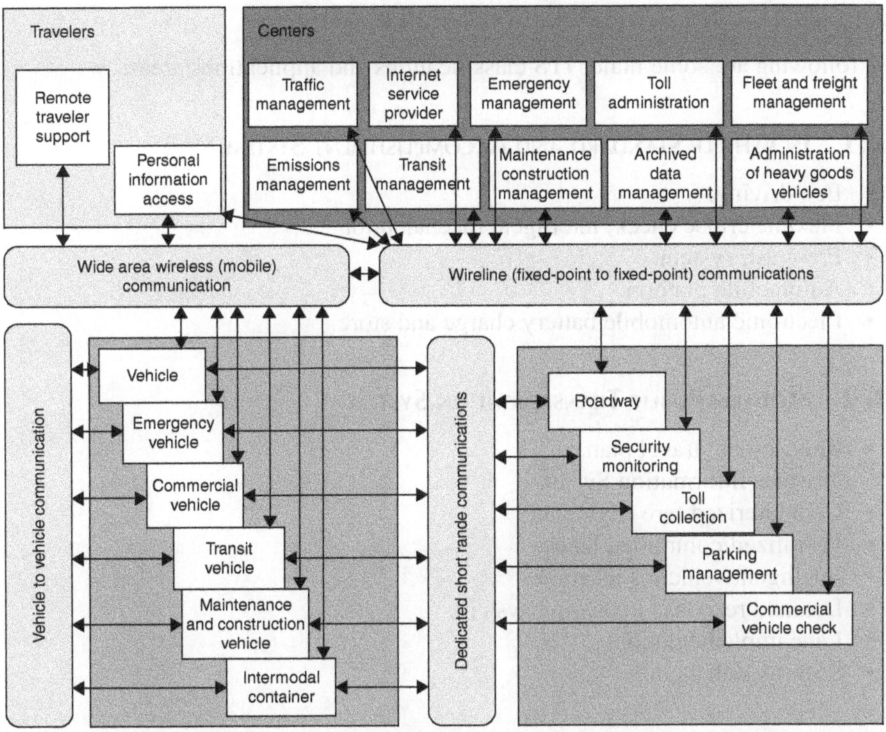

FIGURE 7.2 Physical architecture of ITS.

can aid in the decision-making process when it comes to transportation networks and their efficiency [26–28]. Figure 7.2 shows the physical architecture of Intelligent Transport System.

7.5 EFFECTIVE PARAMETERS TO SUPPORT ITS UTILIZATION

The following are the effective parameters to support ITS utilization:

- Good ability to understand ITS advantages.
- To increase research on ITS utilizations.
- To establish a national ITS standard.
- Legal structure for ITS uses.
- National government's steps to invest in ITS.
- Public and private collaboration to support ITS.
- Need to establish a national organization for ITS.
- Strengthening cooperation with others.
- Improved staff skills.
- Improved capacity.
- All other responses.

7.6 ITS CLASSIFICATIONS AND APPLICATIONS

The following are some major ITS classifications and applications areas:

7.6.1 In Vehicle Standard and Accomplishment System

- Eco-driving.
- Flexible cruise check, intelligent race adaption.
- Pre-crash system.
- Automobile platoon.
- Electronic automobile battery charge and store.

7.6.2 Modern Public Transportation System

- Multi-modal travel plan.
- Traveler Information System.
- Computerized fare payments.
- Prioritized/committed lanes.
- Intelligent vehicle's location.
- Demand response managing system.
- Lane implementation.
- Security enhancing.

7.6.3 Traveler Information System

- Modern traveler information system.
- Route guiding navigation system.

7.6.4 ICT Systems and Platforms

- Smart card payment.
- E-mail, cloud computing and central databases.
- Video-conferencing.
- Online shop and servicing.
- Vehicle allocation scheme.
- Social networking.

7.6.5 Transportation Network Management System

- Parking database.
- To collect and monitor info.
- Smart charging.
- City traffic controlling.
- Varying message sign.
- Access control.
- Speed limiting.

- Lane control.
- ANP recognition.

7.7 STRATEGIC POLICY AND TECHNOLOGY SOLUTIONS

While demonstrable proof of low-carbon benefits from ITS plans is growing, ITS systems must be integrated into nationwide, regional, and localized strategies in order for the advantages to be realized. National authorities can adopt national effectiveness regulations and implement them; adopt carbon-reduction-friendly fiscal policies; promote and fund less pollute transportation projects; support programmes to improve education as well as understanding; and draught climate concerned international agreements. An understanding of the political climate thus provides a valuable qualitative framework in which the proclivity for prioritizing ITS may be investigated using a more analytical manner [29–32].

Such policies are expected to differ internationally, depending on parameters such as quality of economic growth as well as the maturing of the transportation mechanism, and under different national and regional governance systems. Essentially, though, understanding why and how innovations have been prioritized in previous policy choices is critical to predicting future predisposition to emphasize technological solutions [33–34]. A contrast is made here between adaptation and mitigation strategies from the perspective of environmental goals.

Changing climate may have an increase in the frequency, but not always seriousness, of accidents in metropolitan areas, as well as an increased occurrence as well as degree of traffic congestion. More is being discussed about metropolitan regions' growing influence in changing climate governance including their expanding ability to initiate sustainable transitions [35–38]. When transportation policies are implemented at the city-region level, they can acquire flexibility and so respond more effectively to changing conditions.

The following are the key obstacles to realizing the advantages of ICT enhanced transportation strategies:

- An awareness of the long-term planned importance of economic as well as environmental sustainability, especially in the aftermath of the financial crisis. ITS that can prove both performance and environmental improvements would be welcomed by a combined prioritizing.
- In perspective of vertical divisions of devolution and centralization, a governance system that is adequately flexible. This is required to sustain major investment as well as maintenance costs resolutions, which may change in proportion among various categories as well as arrangements of ICT advanced transportation mechanisms.
- Horizontal (cross-sectorial) strategies with a suitably fluid interface that enables comprehensive finding of and financing backing for ICT related actions that have an effect outside of the transportation part.

7.8 ANALYSES OF TRANSPORT NETWORKS

For numerous years, network analysis approaches have been used to study social systems, as well as neuron systems, ecological systems, and computer networks. This happened as a result of the methodologies' extensive mathematical learning. The prospects for conducting analysis using graph theory are indeed discussed [39–43]. The very same techniques and coefficients that sociologists have effectively used to evaluate social systems may now be utilized to analyse transportation networks.

Three measurements can be used to study networks (social networks, transport systems, etc.) and are adequate for evaluating their specific properties and the "worthiness" of the overall network. The plurality of the metrics used and computations completed, result in information that functions as the network's "leader," acting as a focal point for the investigated network.

Every network can be explained as a set of vertices and their interconnections:

$$G = (V, E) \tag{7.1}$$

Here:

V – a set of vertices.

E – a set of their interconnections.

The below inter-relation applicable to each network analysis:

$$|V| = N, |E| = M \tag{7.2}$$

The following coefficients are applicable commonly in analyzing:

Normalize degree (ND_i) for an i-th vertex:

$$NDi = \frac{k_i}{N-1} \tag{7.3}$$

Here:

ki – The count of vertex interconnections with other vertices.

N – count of vertices in the network.

The greater the NDi coefficient amount for this i-th vertex, the more valuable this vertex plays in the network.

Eccentricity (E_i) of an i-th vertex in the network:

$$E_i = \max dij \tag{7.4}$$

$$J \varepsilon V$$

Here:

dij – count of links between the vertices, where a link is considered as the minimum distance among vertex i and vertex j.

The lower the E_i coefficient amount for this i-th vertex, the most considerable this vertex acts in the network.

Radius (R_i) of this i-th vertex in the network:

$$Ri = \frac{1}{E_i} \tag{7.5}$$

Closeness (C_i) is given by:

$$Ci = \frac{N-1}{\sum_{j \in V} d_{ij}} \tag{7.6}$$

Here:
N – count of vertices in the network.
dij – count of links between the vertices, where a link is considered as the minimum distance among vertex i and vertex j.

Betweenness (B_i) for an i-th vertex in the network:

$$Bi = \sum_{l \in V} \sum_{k \neq l \in V} \frac{p_{l,i,k}}{p_{l,k}} \tag{7.7}$$

Here:
$p_{l,i,k}$ – count of connections with the lower count of vertices among vertices l and k.
$p_{l,k}$ – count of connections with the lower count of vertices among vertices l and k.

The more the B_i coefficient amount for this i-th vertex, the valuable role this vertex acts in the network.

Clusterization (G_i) of this i-th vertex in the network:

$$Gi = \frac{2L_i}{k_i\left(k_i - 1\right)} \tag{7.8}$$

Here:
Li – count of links between vertices those are situated nearest to the i-th vertex.
ki – count of vertex interconnections to other vertices.

The more the G_i coefficient amount for this i-th vertex, the more valuable role this vertex acts in the network.

Average shortest paths length (L_{ASP}):

$$LASP = \frac{1}{N\left(N-1\right)} \sum_{i \neq j \in V} d_{ij} \tag{7.9}$$

Here:
N – count of vertices in the network.
dij – count of links among the vertices, where a link is considered as the shortest distance among vertex i and vertex j.

Diameter (D) is given by:

$$D = \max_{i \varepsilon V} E_i \tag{7.10}$$

The lower the diameter, the better the network.
Radius of a network (R_N):

$$R_N = \min_{i \varepsilon V} E_i \tag{7.11}$$

The lower the radius of the network, the better the network.
Average node degree (k):

$$k = \frac{1}{N}\sum_{i \in V} k_i \tag{7.12}$$

Here:
N – count of vertices in the network.
ki – degree of i-th vertex in the network.

The more the AND in the network, the better the network.

There are no occasions in either of the existing systems where all connections use the same "significance" rating. Finding and locating such connections can assist in considering the current state of the network and generating suggestions for its enhancement.

7.9 CONCLUSION

The purpose of this chapter was to finer explain the tendency for low pollution advantages from ITS implementation by discussing research issues: first, scientific proof carries the idea that ITS approaches can be applied as well as functioned in really ways of generating environmental advantages; and second, as to if policy goals among national and international stakeholders represent a tendency for ITS implementation in order to give those advantages.

In accordance with the first study question, a study has shown how ITS technologies can give the technical means to increase vehicle economy and current transportation infrastructure while also supporting behavior changes. Remote communication enabled by ICT can decrease this need carbon concentrated physical transportation. Even if it is inevitable, ITS can lower the carbon output of the managing range. The creation of likelihood frameworks for the prioritizing of ITS that fit with in a larger historical area of technology plans in national, regional, and local planning, has answered the second research topic.

The accomplishment of any prospective ITS effects is founded on the idea that ITS strategies have already been prioritized in national and regional planning, and hence engage with a range of stakeholders, ranging from policymakers to local transportation and environmental suppliers. It could be argued that if the ecological and financial ability of ITS is not recognized, there would be less of a due consideration of the advantages when assessing ITS and different strategies, that the ramp-up of ITS will be sluggish and much more disparate, and that investment and financial choices on huge strategies will be less thorough.

The findings highlight the need for such a greater knowledge of ITS advantages, as well as the importance of expressing cross-sectorial linkages (in terms of compensation and services). Considering its opportunity to make considerable improvements in traffic safety and economics, ITS appears to have strong cross-sectorial effectiveness.

Therefore, it is exceedingly improbable that ITS integration in its entirety will be adequate for successful adaptation to climate change in the transport industry. Although technological progress has been and will continue to be critical for the development of a sustainable transportation approach, it is uncertain to include a magic solution. A policy mix which stimulates behavior changes through taxes, carbon price, as well as regulations and a highly sustainable solution to urban planning, automobile and public infrastructure, will be necessary.

7.10 FUTURE SCOPE

The following are the future research scope issues for ITS approach:

- The human-technology interaction.
- Modern technologies in the future.
- Investigation of associated social and economic issues.
- Better demonstration of ITS advantages.
- Issues of ITS implementation/execution.
- Bringing together current technology.

REFERENCES

[1] Y. Tyrinopoulos, and C. Antoniou, Factors affecting modal choice in urban mobility. *European Transport Research Review* 5 (2013) 27.

[2] I. Kłosok-Bazan, and J. Machnik-Słomka, Meaning of sustainable development in innovation management in enterprises. *Economic and Environmental Studies* 17 (2017) 515–525.

[3] P. Sobczak, *Realization of Bus Connections on the Route Sosnowiec-Chorzów as Part of Sustainable Transport in the Silesian Province; Czasopismo Logistyka: Pozna´ n*, Poland (2015).

[4] E. Stawiarska, Cooperation companies in the region of Silesia and Małopolska for the development of smart specialization Transportation and logistics aspects. *Gospodarka Materiałowa I Logistyka.* 12 (2016) 26–32.
[5] S. Dunn, and S. Wilkinson, Hazard tolerance of spatially distributed complex networks. *Reliability Engineering and System Safety* 157 (2017) 1–12.
[6] I. Spyropoulou, M. Penttinen, M. Karlaftis, T. Vaa, and J. Golias, ITS solutions and accident risks: Prospective and limitations. *Transport Reviews* 28 (5) (2008) 549–572.
[7] O. M. J. Carsten, and F.N. Tate, Intelligent speed adaptation: accident savings and cost benefit analysis. *Accident Analysis and Prevention* 37 (2005) 407–416.
[8] J. Schot, and F. Geels, Strategic niche management and sustainable innovation journeys: theory, findings, research agenda, and policy. *Technology Analysis and Strategic Management* 20 (5) (2008) 537–554.
[9] M. R. Martínez-Torres, M. C. Díaz-Fernández, S. L. Toral, and F. J. Barrero, Identification of new added value services on intelligent transportation systems. *Behaviour & Information Technology* (2011).
[10] G. Marsden, and T. Rye, The governance of transport and climate change, *Journal of Transport Geography* 18 (6) (2010) 669–678.
[11] A. Hull, Policy integration: What will it take to achieve more sustainable transport solutions in cities? *Transport Policy* 15 (2008) 94–103.
[12] A. Jonas, and K. Ward, Introduction to a debate on city-regions: New geographies of governance, democracy and social reproduction. *International Journal of Urban and Regional Research* 31 (1) (2007) 169–178.
[13] M. Hodson, and S. Marvin, Can cities shape socio-technical transitions and how would we know if they were? *Research Policy* 39 (2010) 477–485.
[14] M. Betsill, and H. Bulkeley, Looking back and thinking ahead: A decade of cities and climate change research. *Local Environment* 12 (5) (2007) 447–456.
[15] D. Jaroszweski, L. Chapman, and J. Petts, Assessing the potential impact of climate change on transportation: The need for a interdisciplinary approach, *Journal of Transport Geography* 18 (2010) 331–335.
[16] M. J. Koetse, and P. Rietveld, The impact of climate change and weather on transport: An overview of empirical findings. *Transportation Research Part* D 14 (2009) 205–221.
[17] R. Smokers, R. Vermeulen, R. van Mieghem, R. Gense, I. Skinner, and M. Fergusson et al., Review and analysis of the reduction potential and costs of technological and other measures to reduce CO2-emissions from passenger cars, Final report on behalf of the on behalf of the European Commission (DG-ENTR) (2006).
[18] B. Pel, and F.A. Boons, Transition through subsystem innovation? The case of traffic management. *Technological Forecasting and Social Change* 77 (8) (2010) 1249–1259.
[19] M. Barth, and K. Boriboonsomsin, Energy and emissions impacts of a freeway-based dynamic eco-driving system. *Transportation Research Part D* 14 (6) (2009) 400–410.
[20] J. Pol, ITS technology adoption and observed market trends from ITS deployment tracking Final Report, ITS Joint Program Office, Research and Innovative Technology Administration, U.S. Department of Transportation (2010).
[21] M. Jänicke, and S. Lindemann, Governing environmental innovations. *Environmental Politics* 19 (1) (2010) 127–141.
[22] S. Mandell, Policies towards a more efficient car fleet. *Energy Policy* 37 (2009) 5184–5191
[23] T. Tsekeris, and S. Voß, Design and evaluation of road pricing: state-of-the-art and methodological advances. *Netnomics* 10 (2009) 5–52.

[24] R. Gross, P. Heptonstall, J. Anable, and P. Greenacre, What policies are effective at reducing carbon emissions from surface passenger transport? A review of interventions to encourage behavioural and technological change. UK Energy Research Centre (2019).

[25] G. Santos, H. Behrendt, and A. Teytelboym, Part II: Policy instruments for sustainable road transport. *Research in Transportation Economics* 28 (2010) 46–91.

[26] J. N. Barkenbus, Eco-driving: An overlooked climate change initiative. *Energy Policy* 38 (2010) 762–769.

[27] W. Black, and M. van Geenhuizen, ICT innovation and sustainability of the transport sector. *European Journal of Transport and Infrastructure Research* 6 (1) (2006) 39–60.

[28] H. Roby, Workplace travel plans: past, present and future. *Journal of Transport Geography* 18 (2010) 23–30.

[29] S. Cairns, L. Sloman, C. Newson, J. Anable, A. Kirkbride, and P. Goodwin, *Smarter choices changing the way we travel*. London: Department for Transport (2021).

[30] C. Perez, The double bubble at the turn of the century: technological roots and structural implications. *Cambridge Journal of Economics* 33 (2020) 779–805.

[31] R. Argiolu, R. van der Heijden, I. Bos, and V. Marchau, Intelligent transport systems and preferences for office locations. *Environment and Planning A* 40 (2008) 1744–1759.

[32] G.A. Giannopoulos, The application of information and communication technologies in transport. *European Journal of Operational Research* 152 (2019) 302–320.

[33] A. Curry, T. Hodgson, R. Kelnar, and A. Wilson, Intelligent Infrastructure Futures: The Scenarios Toward 2055. Foresight Programme of the Office of Science and Technology, Project on Intelligent Infrastructure Systems (2018).

[34] G. Cohen, I. Salomon, and P. Nijkamp, Information-communications technologies (ICT) and transport: does knowledge underpin policy. *Telecommunications Policy* 26 (2020) 31–52.

[35] F. W. Geels, Processes and patterns in transitions and system innovations: Refining the coevolutionary multi-level perspective. *Technological Forecasting and Social Change* 72 (2018) 681–696.

[36] B. Elzen, F. W. Geels, and K. Green (Eds.), *System innovation and the transition to sustainability: theory, evidence and policy*. Cheltenham: Edward Elgar, (2019).

[37] International Panel on Climate Change, Fourth Assessment Report on Climate Change. Geneva, (2017).

[38] L. Chapman, Transport and climate change: a review. *Journal of Transport Geography* 15 (2017) 354–367.

[39] D. Banister, and D. Stead, Impact of information and communications technology on transport. *Transport Reviews*, 24 (5) (2018) 611–632.

[40] A. P. J. Mol, *Environmental reform in the information age: The contours of informational governance*. Cambridge: Cambridge University Press, 2021.

[41] G. P. J. Verbong, and F.W. Geels, Exploring sustainability transitions in the electricity, sector with socio-technical pathways. *Technological Forecasting and Social Change*, 77 (8) (2018) 1214–1221.

[42] F. W. Geels, Technological transitions as evolutionary reconfiguration processes: a multilevel perspective and a case-study. *Research Policy* 31 (8) (2021) 1257–1274.

[43] J. Schot, and F. Geels, Strategic niche management and sustainable innovation journeys: theory, findings, research agenda, and policy. *Technology Analysis and Strategic Management*, 20 (5) (2019) 537–554.

8 An Application Oriented Integrated Unequal Clustering Algorithm for Wireless Sensor Network

G. Vennira Selvi, A. Bharathi Anbarasan, B. Akora Murthy, and S. Prabavathy

8.1 INTRODUCTION

WSN includes numerous practical applications inside the medical services, for example estimating biometric parameters of patients. Clustering could provide an energy efficient framework for wireless sensor networks [1]. Intuitively, the size of clusters varies according to different applications [2, 3]. It is essential to provide a general clustering framework which could provide tailored clusters to diverse applications. In order to achieve less energy consumption while communication between nodes, this paper introduces an application-based integrated unequal clustering algorithm for sensor nodes for extending the lifetime of the network.

The main contributions of this chapter are the following:

1. To increase the network lifetime, it constructs an unequal clustering and increases the number of clusters nearer to the BS to mitigate the "hot spot" issue.
2. The frequent re-election of Cluster Head is avoided by assigning predefined threshold energy level for each sensor node.
3. The hot spot problem appears when employing the multi-hop communication in clustering algorithms.
4. The proposed framework argues both inter-cluster communication and intra-cluster communication among the nodes.
5. It also considers the re-election of Cluster Heads and residual energy are not adequate to balance the energy consumption between the entire network. To

DOI: 10.1201/9781003320074-8

address the problem, unequal clustering algorithms are introduced to balance the energy consumption among the entire network.

6. Using this framework, suitable algorithm for a specific sensing field can be elected manually.

8.2 RELATED WORK

Energy Efficient Level-based and Time-based clustering (LTCA) [4] algorithm construct clusters with unequal size associated to the boundaries between upper and lower energy level. It partitions the network into clusters according to the energy level of each node. The energy level is calculated by taking the difference between energy utilization of each Cluster Head in both single-hop and multi-hop communication models. Based on the saved energy, few clusters enlarge their sizes to occupy the remaining nodes for an efficient communication.

In Energy Efficient Heterogeneous Clustered Scheme for heterogeneous WSNs (EEHC) [5], weighted probability of each node is determined to elect the Cluster Heads (CH) in EEHC. The CMs can directly communicate with selected CH. CHs aggregate the data from its Cluster Members (CM) and transfer it to the Base Station (BS). Three different types of nodes with different thresholds are used for each type. It also guarantees that all three types of nodes become CH depending upon their probability of weight. Due to the variable cluster count, EEHC show the way to unequal sized cluster.

A Probability Driven Unequal Clustering Mechanism for WSNs (PRODUCE) [6] composes the network in the form of clusters with unequal in size. The various types of probability are assigned to elect the Cluster Head. In PRODUCE, the clusters nearer to the Base Station are smaller in size and it increases the size of clusters farther away from Base Station. The PRODUCE prompts unequal sized cluster because of variable cluster count but it spends more energy in re-clustering phase.

From the survey, it is transparent that the proposed energy efficient clustering algorithms depletes more energy when re-electing the Cluster Head periodically and collecting data from its CMs. Few algorithms provide the resolution for mitigating the hotspot problem and few are focused to maximize the effectiveness of network lifetime but both the problems are not mentioned together.

8.3 A FRAMEWORK OF UNEQUAL CLUSTERING ALGORITHM FOR WIRELESS SENSOR NETWORKS (FUCA)

The WSN is homogeneous, the network has same residual energy and communication range of sensor nodes [7]. Each node is stationary, after deployment in sensing field. Unique ID is assigned to all the nodes. Each link is symmetric. No local information and synchronization is required on each node. No synchronization service is required. The assumptions guarantee that a network structure can be represented by a graph which helps to analyze the proposed work. The process of a network begins with set-up of the clusters and data transmission. In set up of the clusters, the sensor nodes are grouped into clusters and the sensed information is transferred in data transmission phase.

The framework of Unequal Clustering algorithm uses different user parameters to achieve the goal of prolonging lifetime of the network by extending the lifetime of each sensor node in Wireless Sensor Network. This framework is applicable for both indoor applications such as home automation and outdoor applications such as military applications, weather report generation, etc. The framework shown in Figure 8.1 consists of two components: Node Elector and Energy Trust Value which are used to elect and re-elect the CH.

The node elector elects the CH according to the energy level of each sensors and the number of neighbors. Once the CH is initialized, it calculates the competition radius for constructing unequal clusters to mitigate the hotspot issue. The energy trust value is responsible for re-electing the Cluster Head which is applied to the threshold energy function. Threshold energy function invokes the re-clustering algorithm if it satisfies the specified constraint. The appropriate proposed algorithms and parameters are chosen according to the applications of WSN. The components of the framework are decribed in detail below.

The framework includes the proposed algorithms namely BUCA [8], UCAPN [9], FPUCA [10], MinD-CDS [11] and OUCA [12] for large scale WSN. FUCA selects the best algorithm based on sensing field and the observed parameters such as energy depletion, network lifespan and communication overhead. The existing frameworks considers a specific protocol for transferring data over multi-hop communication and do not integrate with any other protocols. In the existing framework, a large number of communicating nodes participate for transferring data which results a communication overhead and unbalanced energy consumption among sensor nodes. Therefore, the intent of this framework is to integrate the proposed algorithm together to select the suitable algorithm for gathering the data in specified sensing areas. The proposed framework is a generic framework in which the proposed algorithms are combined together to provide an energy effectual unequal clustering technique for various applications in WSN.

The framework of Unequal Clustering algorithm uses different user parameters to achieve the goal of prolonging lifetime of the network by extending the lifetime of each sensor in Wireless Sensor Network. This framework is applicable for both indoor applications such as home automation and outdoor applications such as military applications, weather report generation etc. The framework shown in Figure 8.1 contains two components: Node Elector and Energy Trust Value which are used to elect and re-elect the CH. The node elector elects the Cluster Head according to the energy level of each sensor and number of neighbors. Once it is initialized, it calculates the competition radius for constructing unequal clusters to mitigate hot spot issue. The energy trust value is responsible for re-electing the Cluster Head which is applied to the threshold energy function. Threshold energy function invoke the re-clustering algorithm if it satisfies the specified constraint. The appropriate proposed algorithms and parameters are chosen according to the applications of WSN. The components of the framework are described in detail below.

8.3.1 Neighbor Node

In neighbor node, every node sends a beacon signal to the node neighbor to observe the number of neighbors and its residual energy. The notation of neighbor node observer is assumed as follows:

Notation 1: For a node 'i', a neighbor node of 'i' is a node that is reachable from i with one-hop wireless communication.

To monitor the energy level of each node, the following assumption should be made to satisfy the energy prediction model in WSN. The sensor nodes are distributed randomly. After deployment, the nodes are stationary. The unique address ID is allotted to every node. The data from sensing environment are collected periodically and forwarded to the BS. The sensor nodes are monitored at discrete intervals of time. Every node calculates the space between the nodes according to the strength of the received signal. The BS is placed at the middle of the sensing area and it is controlled by the wireless transmission control. The proposed algorithm uses the radio dissipation of energy method to transmit the amount of data along a distance d and is defined as follows:

$$ET_x(k,d)=\begin{cases} k*E_{elec}+k*E_{fs}*\mathrm{d}^2, & d<\mathrm{d}_0 \\ k*E_{elec}+k*E_{fs}*\mathrm{d}^4, & d\geq \mathrm{d}_0 \end{cases} \cdots \tag{8.1}$$

Where, E_{elec} represents the electronic energy, E_{fs} represents amplifier energy d is the difference between receiver and transmitter and d_0 is threshold distance. If d is less than the threshold d_0 then d^2 is single-hop communication otherwise d^4 is multi-hop communication. The energy spent for receiving k bits data from its neighbor is calculated using the following formula:

$$ER_x(k) = k \times E_{elec} \tag{8.2}$$

Where ER_x is the amount of energy depletes it energy to receive k bits of data.

8.3.2 Energy Level

The Energy Level continuously monitors the energy level of every node by transferring a beacon signal to its nearest neighbor. The average energy level of each node is computed using equation (8.3).

$$E_a=\frac{1}{d}\sum_{i=1}^{d} s_i.E_r \tag{8.3}$$

where d is the number of neighborhood and E_r is residual energy of specified node s_i. The waiting time is calculated for sending a message to its neighbor is calculated as follows.

TABLE 8.1
Nearest Neighbor Node Information

Node ID	Residual Energy (E) in Joules	Node Status	Number of Neighbors Nodes
N1	2.0	Unknown	5
N2	2.8	Unknown	8
.	.	.	.
.	.	.	.

$$t = \left[(1-\propto) * \frac{N_i}{N_{max}} \right] + \left[\propto * V_{random} \right] \tag{8.4}$$

Where N_i is the number of neighborhood, N_{max} is the total amount of nodes, α is a constant coefficient among 0 and 1, V_{random} is a random number between 0 and 1. For the duration of round 0, every node has equal energy level to further nodes. The neighbor node information collected from each node is shown in Table 8.1.

The Node ID represents the unique ID for each node i.e., N1, N2, N3,……… …………,Nn where n is number of nodes present in the network. Residual energy (E) represents the energy level of each node that will be measured in joule. The status of node is mentioned in the column of Node Status and the number of neighbors for every node is specified in the column of Number of Neighbors Nodes.

The energy level manager helps to elect the Cluster Head. The maximum residual energy among its neighbor node is elected as a CH.

8.3.3 Competition Radius

To avoid hot spot problem, the proposed algorithms should form more clusters closer to the BS. To construct clusters with unequal in size and every node is responsible to calculate its individual competition radius R_c using equation (8.5).

$$R_C = 1 - c\frac{d_{max} - d(s_i, BS)}{d_{max} - d_{min}} R_o c \tag{8.5}$$

The distance between maximum nodes and minimum nodes and Base Station is represented as d_{max}, d_{min}. The weighted factor specified as c and maximum competition radius represented as $R_o c$. Every node elects the nearby CH and conveys the Join_Msg with its ID and residual energy. Every CH prepares the node schedule list for each node based on expected Join_Msgs and transmits the list to its CMs.

The algorithm for selecting a CH is given in Algorithm 1:

Algorithm 1: Unequal cluster formation

```
Begin
status=unknown
t=waiting time to broadcast CH message
r=radio signal
while (status==unknown) do
broadcast ADV_msg within
compute R_c competition radius
broadcast HELLO_Msg from node i to neighbor node j
collect neighbor node information
update it in neighbor node table
calculate average residual energy
if (E_i>E_j) then
broadcast HEAD_Msg(ID=i, d, E,NN)
exit
if (current time>t) then
if (HEAD_Msg receives from node i from node j) then
status=unknown
else
continue
end if
else
status=header
broadcast HEAD_Msg to its neighbor
send JOIN_Msg from node j to node k
end if
if status==header then
receive JOIN_Msg from its neighbor node
set status=member
end if
```

The aforementioned algorithm selects the Cluster Head according to number of neighbor nodes and its residual energy. It builds clusters with unequal in size by increasing the amount of clusters nearer to the BS to avoid the hot spot issue and to increase the network lifetime.

8.3.4 Trust Value

The Trust Value decides which node will become a CH. Once the CH is elected, it forwards the message to its neighbor. The neighbor node directs a join message to nearby Cluster Head. The neighbor node observer and energy level monitor maintains a table with number of neighbors and its energy level.

8.3.5 Threshold Energy Function

The Threshold Energy Function avoids the frequent re-election of Cluster Head. Once the clusters are constructed, the threshold energy value is assigned to each Cluster Head. The Cluster Head is rotated when its residual energy is lower than the threshold energy value.

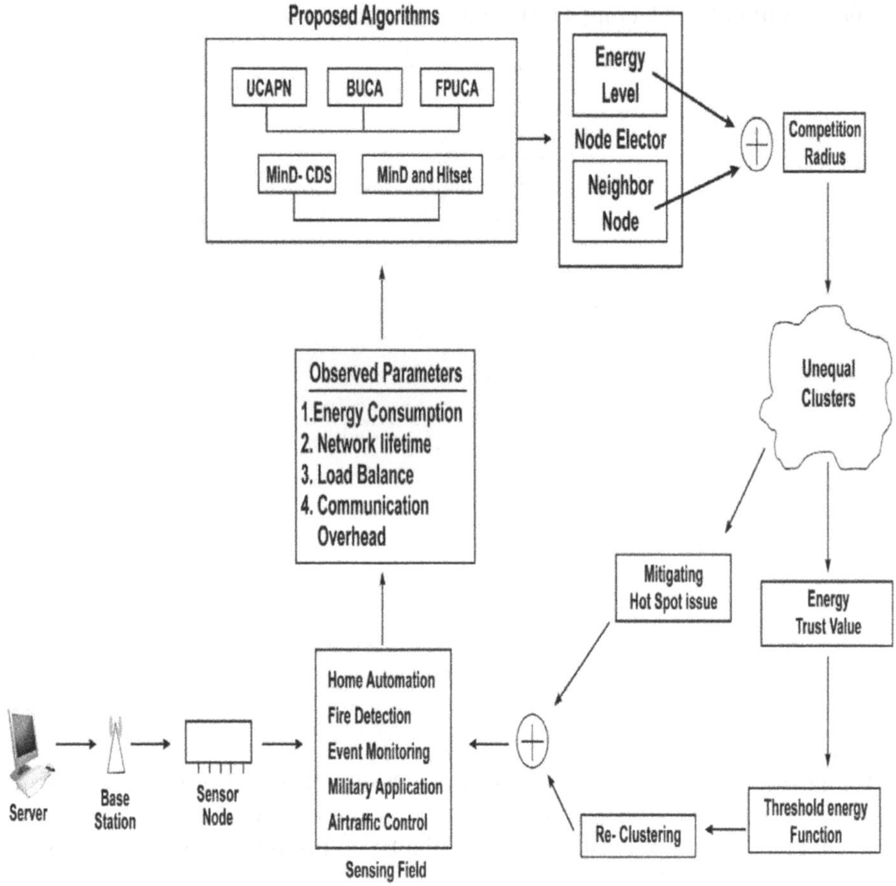

FIGURE 8.1 Framework for unequal clustering algorithm.

8.4 OVERVIEW OF FUCA

FUCA integrates the proposed algorithms together in order to mitigate the hot spot issues and to extend the lifetime of the network in various applications of WSN. In FUCA, the proposed algorithms are chosen based on the observed parameters and nodes distributed in the sensing area. Once the nodes are deployed, the required parameters are selected for analyzing the performance of the specified application in sensing field. The FUCA needs to decide three main things before electing the proposed algorithm. They are, 1) sensing field is large scale or small scale, 2) energy driven or time driven method for rotating CH and 3) Node degree is maximum or minimum. The algorithm is chosen based on the following combination.

- The proposed algorithm UCAPN is elected when the nodes are deployed in small scale and uses time driven method for rotating Cluster Head.

- BUCA is elected when nodes are deployed in small scale and energy driven technique for rotating Cluster Head.
- FPUCA is elected when the nodes are deployed in large scale sensing area and forecasting the energy level of each node in advance at discrete time interval.
- For a large scale WSN, the MinD-CDS algorithm is chosen with minimum node degree and rotating the Cluster Head with energy driven technique.
- The OUCA is chosen when the nodes are in large scale WSN with maximum node degree and energy driven technique.

After electing the proposed algorithm, the node elects the Cluster Head according to the energy level of each node and number of neighbors using node elector. To construct the clusters of unequal size, the competition radius is calculated for each sensor node. The proposed algorithms BUCA, MinD-CDS and OUCA use the energy trust value and threshold energy function for minimizing the recurrent re-election of CH which preserves some more energy for prolonging network lifetime. Finally, the processed data are gathered by the CH and onward to the server through the BS.

8.5 APPLICATION OF FUCA IN STRUCTURAL HEALTH MONITORING

In recent years, using Wireless Sensor Networks for Structural Health Monitoring (SHM) has established increased consideration from diverse fields such as civil, mechanical, structural and aeronautical engineering due to low cost, low maintenance, flexibility, fast and easy connection through wired sensor networks [13]. The target of SHM is to monitor complex events such as impairment, crash and erosion in physical system [14]. In the proposed framework, the sensor nodes are distributed on different places of structure to gather the structure response. Therefore, the sensing area of SHM is small scale and the nodes should response only if it finds a problem in the structure. In this situation, the proposed algorithm BUCA is suitable for detecting the location of damaged structure. This data is then transmitted through the Base Station to the destination. In BUCA, the group of sensors detect damage event fully changes over time.

8.6 APPLICATION OF FUCA IN HABITAT MONITORING

In WSN, the Habitat Monitoring necessitates that each node should detect the environmental condition periodically and forward the detected data to the BS [15]. In temperature monitoring application, there is only one report from each sensing area. Hence, one temperature record from each CH is adequate if the communication range of each sensor node located roughly 10 meters to 15 meters with multi-hop communication [16]. The habitat monitoring is a large scale WSN and difficult to recharge in sensing field. In habitat monitoring, each node sends the collected data to the BS if any unexpected even occur in sensing field. Therefore, it is necessary to elect the optimized CH for monitoring the node to a longer life. The optimized cluster head algorithm is chosen for detecting the event in an unattended environment. In OUCA, the clusters are constructed evenly over the sensing area. It also reduces the amount of clusters to attain the energy efficiently and increases the rate of data transmission for each cluster member.

8.7 APPLICATION OF FUCA IN HOME CONTROL

In WSN, the application of home control is a good illustration to provide control as well as protection at home [17]. The UCAPN algorithm is suitable for the application of home control which provides the detail data about electric, gas and water usage by distributing the sensor network in the specified area at home. The proposed algorithm also provides the automatic announcement upon discovery of some unused activities in the home.

8.8 SIMULATION AND RESULTS

The performance of the proposed framework is evaluated by conducting a series of simulations. In the experiment, a network of 50–250 sensors are consistently distributed in square field with the radius 200 m *200 m. After the formation, nodes become static, no longer moving. Assuming the Base Station is placed at the center of the sensing area; experiment is carried out on the factor of energy depletion, network lifetime and average delay time.

8.8.1 Scenario-1 – Performance Evaluation of FUCA in Habitat Monitoring

The proposed framework is analyzed itself to select the best proposed algorithm for specified applications. In this scenario, the nodes are distributed in the sensing field of monitoring the habitat of weather based on the temperature and humidity. The result of the proposed framework is analyzed according to the performance metrics of energy consumption, network lifetime, cluster formation and amount of clusters formed closer to the Base Station. In this scenario, there are 1000 nodes distributed in the sensing arena with the initial energy of 5 J.

8.8.1.1 Experiment 1 – Performance Evaluation of FUCA Based on Energy Consumption

The energy consumption of the proposed algorithms is analyzed over time. The Figure 8.2 show that the OUCA consumes less energy related than the other proposed algorithms MinD-CDS, FPUCA, BUCA and UCAPN. The proposed algorithm UCAPN consumes 19% of more energy than the proposed algorithm OUCA and BUCA consumes more than 12% energy than OUCA.

In UCAPN, the CHs are re-selected periodically which will consume extra energy. The FPUCA spent some more energy for forecasting the energy level of each sensor node at a discrete interval of time. In MinD-CDS, more communicating nodes are participated to transfer a data by constructing clusters with smaller in size using CDS.

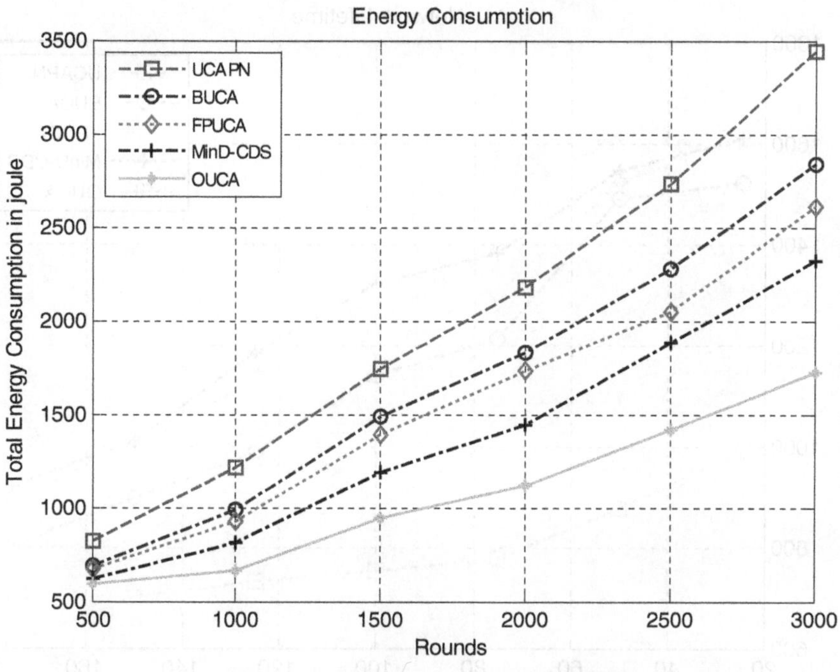

FIGURE 8.2 Energy consumption in the sensing area of habitat monitoring.

TABLE 8.2
Energy Consumption in the Sensing Area of Habitat Monitoring, Number of Rounds vs. Total Energy Consumption in Joules

Rounds in Sec.	UCAPN	BUCA	FPUCA	MinD-CDS	OUCA
500	822	692	671	616	593
1000	1211	983	923	813	662
1500	1745	1483	1391	1187	947
2000	2184	1826	1736	1446	1120
2500	2733	2273	2048	1882	1414
3000	3440	2841	2612	2321	1726

But in OUCA, the Hitting Set is applied over the dominating set to elect the optimized CH and to reduce the amount of communicating nodes for data transfer. Therefore, the proposed algorithm OUCA consumes less than 6% of energy than MinD-CDS.

8.8.1.2 Experiment 2 – Performance Evaluation of FUCA Based on Network Lifetime

The network lifetime is analyzed by varying the distance from Base Station along with the time. Figure 8.3 shows that the OUCA extends the lifetime of the network by

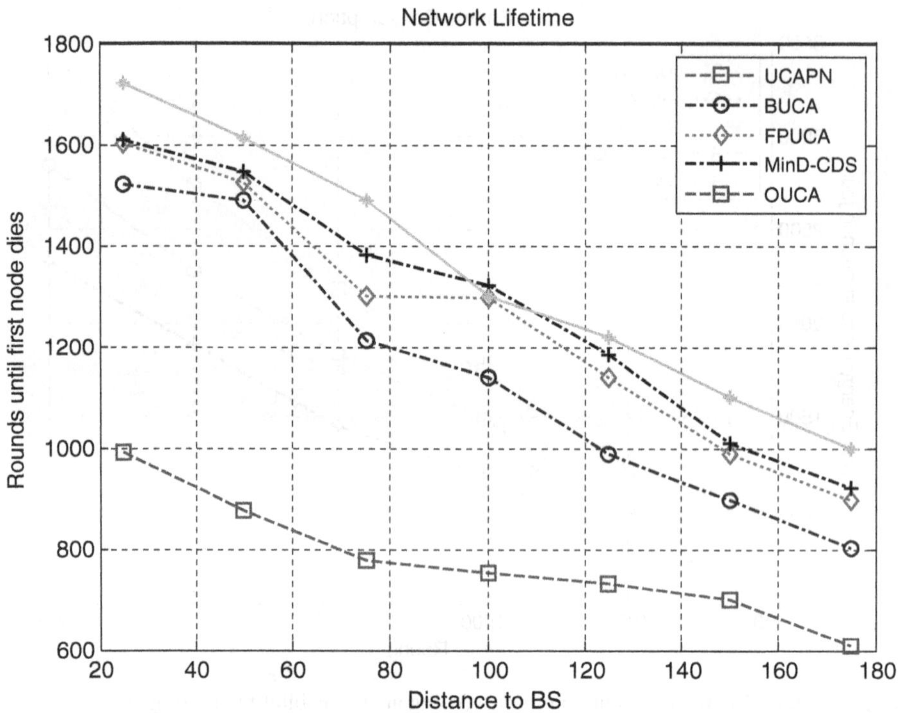

FIGURE 8.3 Network lifetime for habitat monitoring.

TABLE 8.3
Network Lifetime: Distance to the Base Station vs. Time for First Node Dies

Distance to BS	UCAPN	BUCA	FPUCA	MinD-CDS	OUCA
25	990	1520	1600	1608	1720
50	876	1488	1526	1545	1612
75	778	1211	1301	1380	1489
100	753	1140	1298	1320	1301
125	732	989	1140	1183	1221
150	701	896	989	1010	1101
175	609	802	898	921	998

23% over UCAPN, 14% over BUCA, 12% over FPUCA and more than 7% to MinD-CDS. In OUCA, the energy consumption among the sensor nodes are minimized by electing an optimized Cluster Head. Therefore, OUCA extends the lifetime of the network more efficiently than other proposed algorithms.

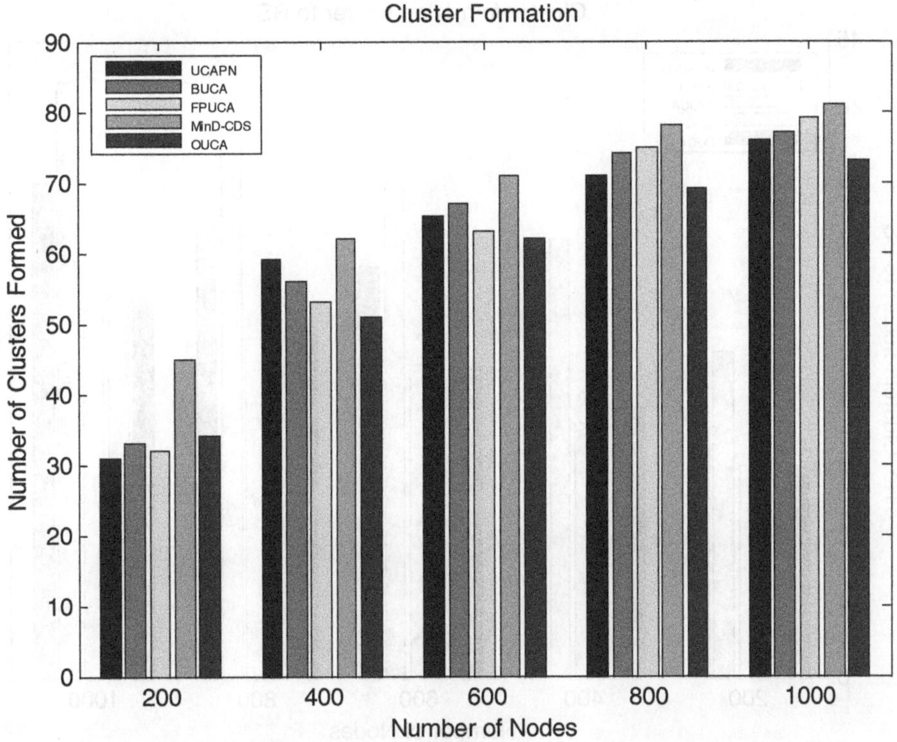

FIGURE 8.4 Cluster formation in habitat monitoring.

8.8.1.3 Experiment 3 – Performance Evaluation of FUCA Based on Cluster Formation

In experiment 3, the performances of the proposed algorithms are analyzed according to the amount of clusters formed by varying the number nodes from 200–1000. In MinD-CDS, the CHs are selected according to the parameters of high remaining energy and minimum node degree. The number of nodes in the cluster is reduced and the numbers of clusters are increased. The proposed algorithms UCAPN, BUCA, FPUCA and OUCA elect the CH according to the maximum remaining energy, distance from BS and number of neighbors. Figure 8.4 depicts, the proposed algorithm MinD-CDS constructs more clusters compared than other proposed algorithms. MinD-CDS construct the virtual backbone to provide an efficient data transfer among the sensor nodes.

8.8.1.4 Experiment 4–Performance Evaluation of FUCA Based Amount of Clusters Closer to the BS

In order to protect from hot spot issues, the proposed algorithms use a competition radius for partitioning the sensing field into layered architecture which helps to construct the unequal size clusters. The layer nearer to the BS has more amounts of

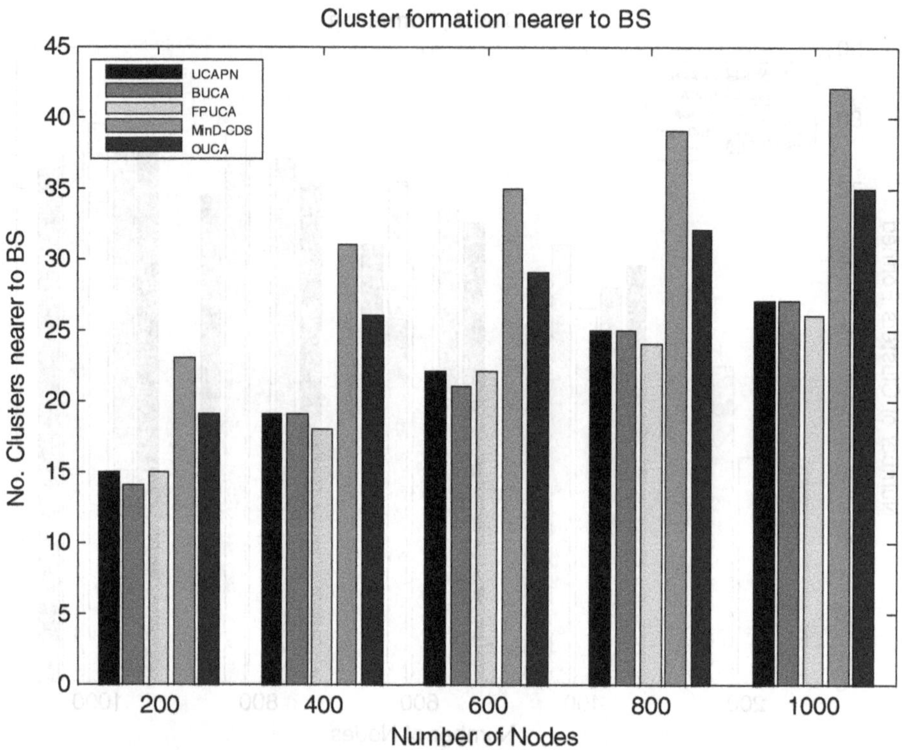

FIGURE 8.5 Clusters nearer to the BS in habitat monitoring.

clusters than farther away from BS. In this framework, the MinD-CDS constructs more amounts of clusters closer to the Base Station compared to other proposed algorithms. MinD-CDS uses the Connected Dominating Set with minimum node degree to select the CH. Hence, it reduces the number of nodes in every cluster and maximizes the amount of clusters especially nearer to the BS. In OUCA, the communicating nodes are reduced by applying the Hitting Set over a dominating set which will result in the number of clusters being reduced compared to MinD-CDS. The MinD-CDS construct 6.2% of clusters more than OUCA. The proposed algorithms UCAPN, BUCA and FPUCA constructs approximately equal number of clusters closer to the BS and construct 13% of clusters less than MinD-CDS.

8.9 CONCLUSION

From scenario 1, the proposed algorithm OUCA consumes less energy and maximizes the network lifetime by reducing the communicating node by electing an optimized Cluster Head. But in cluster formation and in the amount of clusters closer to the BS, the algorithm MinD-CDS constructs 6% of clusters more than OUCA. In clustering, the communicating nodes are responsible for collecting and conveying data to the

Base Station. Therefore, communicating nodes consume more energy than other nodes. The proposed algorithm OUCA decreases the number of communicating nodes and energy consumption is balanced among the sensor nodes. The network lifetime is increased by minimizing the energy depletion of the sensor nodes. Hence it is found that OUCA is suitable for habitat monitoring application in WSN.

REFERENCES

[1] I Dietrich, F Dressler, "On the Lifetime of Wireless Sensor Networks", *ACM Transaction Sensor Networks*, vol. 5, no. 1, pp. 1–5, Jan. 2009.

[2] O Younis, S Fahmy, "Distributed Clustering in Ad Hoc Sensor Networks: A Hybrid, Energy-efficient Approach", *IEEE Transaction Mobile Computing*, vol. 3, no. 4, pp. 366–379, Oct. 2004.

[3] J Yang, D Zhang, "An Energy-balancing Unequal Clustering Protocol for Wireless Sensor Networks", *Information Technology Journal,* vol. 8, no. 1, pp. 57–63, Aug. 2009.

[4] F Tashtarian, M T Honary, A T Haghighat, J Chitizadeh, "A New Energy-Efficient Level-based Clustering Algorithm for Wireless Sensor Networks", *International Conference on Information, Communications & Signal Processing*, vol. 1, no. 1, pp. 1–5, Dec. 2007.

[5] D Kumar, "Energy Efficient Heterogeneous Clustered Scheme for Wireless Sensor Networks", *Computer Communications*, vol. 1, no. 1, pp. 78–83, Aug. 2009.

[6] J-H Kim, C S Hussain, "PRODUCE: A Probability-Driven Unequal Clustering Mechanism for Wireless Sensor Networks", *International Conference on Advanced Information Networking and Applications*, vol. 1, no. 1, pp. 356–345, Mar. 2008.

[7] V Mhatre, C Rosenberg, "Design Guidelines for Wireless Sensor Networks: Communication, Clustering and Aggregation", *Ad Hoc Networks*, vol. 2, no. 1, pp. 45–63, Jan. 2004.

[8] G Vennira Selvi, R Manoharan, "Balanced Unequal Clustering Algorithm for Wireless Sensor Network*", International Journal of Communication and Networking System*, vol. 4, no. 1, pp. 327–332, Jun. 2015.

[9] G Vennira Selvi, R Manoharan, "Unequal Clustering Algorithm for WSN for Prolonging Network Lifetime (UCAPN)", *International Conference on Intelligent Systems, Modelling and Simulation*, vol. 1, no. 1, pp. 456–461, Aug. 2013.

[10] G Vennira Selvi, "Unequal Cluster Based Fault Prediction Algorithm for Wireless Sensor Networks", *International Journal of Convergence Computing,* vol. 1, no. 4, pp. 244–258, Mar. 2015.

[11] G Vennira Selvi, P Udaya kumar, J Senthil Murugan, E Purushotham, S Kokila, "Minimum Degree Connected Dominant Set Unequal Clustering For Wireless Sensor Networks (Mind-CDS)", *Webology*, (ISSN: 1735-188X) vol. 19, no. 2, 2022, pp. 1686–1699.

[12] G Vennira Selvi, V Muthukumaran, A C Kaladevi, S Satheesh Kumar, B Swapna, "Integrated Dominating and Hit Set-inspired Unequal Clustering-based Data Aggregation in Wireless Sensor Networks", *International Journal of Intelligent Computing and Cybernetics*, ISSN: 1756-378X, 16 March 2022.

[13] W Dargie, R Mochaourab, A Schill, Lin, "A Topology Control Protocol Based on Eligibility and Efficiency Metrics", *Journal of Systems and Software*, vol. 84, no. 1, pp. 102–111, Jan. 2011.

[14] A C Khosrowshahi, B Arasteh, S T Afshord, B Arasteh, "A New Strategy for Optimizing Energy and Delay in MCSMAC Protocol", *Indian Journal of Science and Technology*, vol. 7, no. 11, pp. 1756–1764, Nov. 2014.

[15] M Vinothini, A Umamakeswari, "Reliable Data Transmis¬Sion Using Efficient Neighbor Coverage Routing Protocol in Wireless Sensor Network", *Indian Journal of Science and Technology*, vol. 7, no. 12, pp. 2118–2123, Dec. 2014.

[16] Y Liao, H Qi, Li, "Load-balanced Clustering Algorithm with Distributed Self-organization for Wireless Sensor Networks", *Sensors*, vol. 13, no. 5, pp. 1498–1506, May 2013.

[17] J Bahi, A Makhoul, M Medlej, "A Two Tiers Data Aggregation Scheme for Periodic Sensor Networks", *Ad Hoc & Sensor Wireless Networks*, vol. 21, no. 2, pp. 77–100, Mar. 2014.

Index

Note: Page numbers in **bold** refer to tables, and those in *italics* refer to figures.

V

W